Boundaries of the Self

Boundaries of the Self: Gender, Culture and Spaces

Edited by

Debalina Banerjee

CAMBRIDGE SCHOLARS PUBLISHING

Boundaries of the Self: Gender, Culture and Spaces
Edited by Debalina Banerjee

This book first published 2014

Cambridge Scholars Publishing

12 Back Chapman Street, Newcastle upon Tyne, NE6 2XX, UK

British Library Cataloguing in Publication Data
A catalogue record for this book is available from the British Library

ISBN (10): 1-4438-5706-8, ISBN (13): 978-1-4438-5706-2

TABLE OF CONTENTS

PREFACE

This book is an attempt to capture the spirit of the conference on *Women and Spaces: Engendering and Re-gendering Identities* organised by the departments of English, Economics and History, Vidyasagar Evening College, University of Calcutta; in collaboration with Ramakrishna Mission Institute of Culture, Golpark, Kolkata, on the 29th and 30th of August, 2011. The presentation of academic papers, a talk on women and trafficking by the ex-Commissioner of Kolkata Police, a presentation by eminent theatre personality Saoli Mitra on her plays, followed by the screening of a special film on women and trade, *1^3 Three; yet one!* by Pramod Dev (independent director), made it a laudable attempt. The topics ranged from colonial spaces, music and empowerment, photography and visibility, and animation films, together with hardcore feminist theory and feminist interpretations of texts, corporate globalisation and its impact on women, women scroll painters, the sex worker in Bengal, identity issues concerning Bengali Muslim women, the life and times of Kadambini Ganguly (the first woman medical practitioner of Bengal) and class inside the household. While all papers presented at the conference have not been included in this volume, some very interesting additions have been made – from microfinance, advertisement, diaspora, re-conceptualisation of class/gender in terms of textual readings, women's education in nineteenth century Bengal (backed by some rare archival data) to the assessment of educational institutions, namely schools, as modern Foucauldian institutions. The entire initiative of the conference and the book would not have been possible without the unending support of the Principal, Dr Ramswarup Gangopadhay, and my fellow colleagues, Swati Maitra, Debashis Joddar and Suparna Pal, from the departments of History and Economics. Discussing women's issues with them on a hot summer evening, over a hot cup of tea, had resulted in the conference. A special thanks to Debashis Joddar for sticking with me through thick and thin in almost every aspect of the conference, especially in handling the financial burden. My colleagues from various academic disciplines and members of the conference committee – Sarbari Ghosh, Madhumita Dasgupta, Samir Kumar Ghosh, Rahul Mazumder, Tamal Das, Sarajit Sardar, Sandip Mukherjee, Debashis Bhattacharya, Ujjal Chakraborty, Sudarshan Pal, Soukhen Joarder and Sushmita Saha – helped make this event a success. They further made me realise the relevance of bringing out such a book with its remarkably interdisciplinary character.

Swami Sarvabhutananda, Secretary, Ramakrishna Mission Institute of Culture, Golpark, Kolkata, Late Mr Utpal Biswas and other staff of the institute, through their unconditional support, helped in the germination of the idea and this book would not have been possible without them. I am extremely indebted to University Grants Commission for providing necessary and vital financial assistance so that our efforts could see the light of day. I extend my gratitude to Cambridge Scholars Publishing for approaching me and giving me this wonderful opportunity to work with them. And last but not the least, a heartfelt thank you to Tanushree Mukherjee for her commitment, sincerity and astute analysis of the texts, without which this book would not have taken its final shape.

"How would you define a woman?" I had asked, talking to a class at the University of Vienna. Surprisingly, the answers were similar in their configuration of a woman merely as a being with breasts, genitalia and the power to procreate. Such anatomically deductive hypothesis seemed to shrink women into the symbolic washroom icon marked 'Ladies'. The category called 'woman' is diverse in its 'authority of experience', cutting across spaces, cultures, communities, ethnicities, nationalities and so on and so forth. Feminism, in terms of theory and praxis, has long branched out into various schools of thought. Hence, planning a conference on women's studies only threatens to open up divisive histories. Therefore, the right way to debate and discuss 'women' seemed to be within the notion of 'spaces' with their inherent plurality of 'feminine', 'feminist' and female (to borrow Elaine Showalter's terminology) identities.

The workings of patriarchy become markedly visible within the networks of social, cultural, political, historical and economic spaces. Ideology is insular and sectarian. Hence, following any one ideological paradigm to probe into women's issues seems impossible in terms of policy and significance. Interrogating patriarchal politics that is both 'sexual' and 'textual' tends to open up the centres and the margins – the loci of patriarchal positioning/repositioning in society, literature and culture. And as these spaces drift, collide and coalesce, we get representations of women both 'real' and 'imagined'. Spaces are, therefore, intrinsically linked to boundaries. Because gender identity and construction involves remaining 'fenced in' and 'breaking out' in the context of hegemonic power play, it is irrevocably connected to the question of transgression, violation and emancipation. Iris Young observes that "a space seems to surround women in imagination which they are hesitant to move beyond… women's space is not a field in which her bodily intentionality can be freely realised but an enclosure in which she feels herself positioned and by which she is confined." John Berger, in his essay "The Split Woman", assesses the presence of women within Western cultures – existent solely in the restricted space of male desire. Such Panopticism, reminiscent of Foucault's 'Panopticon', is true of all cultures and generates a space where women are constantly under male surveillance vis-à-vis 'gaze'.

The aim of this book is to address the intersections between gender and identity by critically examining female spaces. It has famously been argued that men and women are made in culture. The book seeks to explore how spaces – social, political, literary, cultural and historical – affect the identities of women, whether creative, personal, collective, urban or global. Through the scholarly approaches of the international and national contributors, we wish to probe into these spaces and analyse the problematic of gender identities as they are constructed, reconstructed or deconstructed through processes of appropriation, subversion or signification. This book hopes to provide a significant direction in women's studies through dialogues and discussions on the various facets of 'space' and how, in turn, it generates a rhetoric of agency and power or, again, how it annihilates attempts at emancipation and empowerment.

The main purpose is to further explore the diversity of women's experiences and their contributions across cultures. The book attempts to examine knowledge and practices in the light of gender differences and suggest new ways to "conceptualise the relations between the self and the ever changing global communities". Its interdisciplinary nature, drawing from the humanities, arts and social sciences, will give it a wide readership among students, teachers and researchers. Since women's studies is one of the most sought-after disciplines of the contemporary academic world, this book will generate interest and contribute to the dynamic nature of women's studies research. The title of the book is taken from Dr Usha Bande's stimulating keynote address at the conference. The fluidity of gender identities, as they are performed (courtesy Butler), is best exemplified by the metaphor of space. Going beyond boundaries in terms of gender and culture would effectively highlight the processes of gendering, engendering and re-gendering. I end with an excerpt from an interview of Rituparno Ghosh given to Kaustav Bakshi. Ghosh observes:

> …our identities are subject to the body which again is a boundary… I believe in transcending that boundary… the body is in a state of transition… perennially… so is my identity. Therefore, it is not desirable to identify with a single category. It is, in fact, impossible. Everything is in a state of making… eternally… nothing is ever complete… the same is true of the body and, therefore, identity. It's a continuous process. (http://silhouette-mag.wikidot.com/vol10-3-kaustavbakshi)

CHAPTER ONE

INTRODUCTION

USHA BANDE

At the very outset, I would like to compliment the organisers for the theme of the seminar, **"Women and Spaces: Engendering and Re-gendering Identities"**. Without referring to feminism, the theme covers the whole area that comes under the term patriarchy and its associated web of economic, political, social and religious regulations that enforce the domination of women by men. The theme of the seminar, "Women and Spaces: Engendering and Re-gendering Identities", offers a broad spectrum of women's reality. This single word – feminism – is enough to describe the grand network of oppressive forces, suggesting that male domination stretches across national and cultural boundaries, touching various facets of life. This encompasses a vast range of concepts like gender, identity, spaces, self, revision/re-vision/representation/re-presentation and language. This seminar will certainly be focusing on these aspects, taking into consideration genres like fiction, poetry, drama, autobiographies, films, media, and travelogues.

Women's exclusion from history stems from gender formation. Gender role differentiation is associated with gender differences in behaviour, attitudes, and dispositional traits. This differentiation also leads to gender stereotyping. Engendering suggests empowerment which is a necessary ingredient for challenging and transforming unequal political, economic and social structures. And re-gendering means to review the gender stereotype and correct the picture.

As I sat to write this address for the seminar, a poem read long ago, popped up in my mind. The title is "A Feminist Poem on a Flower" and the line that captivated me was: "A flower has no words/And the words that a woman has are not meant for speaking." Flowers symbolise beauty, delicacy, self-sacrifice and giving. Silence signifies inarticulation and the resultant invisibility, loss of self and loss of identity. The premise of the seminar may not be directly flowers and women, words and silences; but it implies these and a lot many other aspects. The theme of the seminar addresses two categories: Spaces and identities and the third, on which these reflect, is the 'woman'. Woman, as a category, stands bracketed as the 'other' by patriarchal norms and implies existence of boundaries that hem her in from all sides – cultural, moral, societal, familial and linguistic. That is why I titled my address as "Boundaries of the Self: Gender, Culture and Spaces".

Central to this orientation is the question: Is there a woman's space? Is there an identity for a woman distinct from her social-familial identity? Identity as an individual? Don't we remember Mira's mother in Shashi Deshpande's, *The Binding Vine*? She would not show her horoscope to the astrologer on the plea that her welfare lies in the welfare of her children and husband. I remember a story titled, "Mera Ghar Kahan" (Where is My Home?), in which a young girl, dislodged from various houses where she works as a domestic help, wonders in the end, 'Mera ghar kahan?'. Do women have a home which they can call their own? These are some of the questions problematised by feminists to uncover women's exclusion from home. Let us turn to a few lines of a poem by Anamika:

Ram, go to school!
Radha, cook a meal!
Ram, come, have a sugar cake!
Radha, sweep the house!
Your brother will sleep now, go and make his bed!
Wonderful! A new home!
Ram, look, this is your room!
And mine? Silly girl,
Girls are air, sunshine, earth
They do not have a home!
Those who have no home
What place can they have?

Space signifies existence of an identity and identity requires space. This proposition has been beautifully revealed by a poet unable to find a space to sit and write:

"Hubby's room/has no entry
For even a beetle.
No disturbance please!
In children's room/No entry for silence."

So what does a woman writer do? Sit in the kitchen. "Kitchen is an island/unto itself", reiterates the above poem. This is not a new situation. Just think, where did the famous novel, *Uncle Tom's Cabin*, get written? At the dining table, literally. Nearer home, in a short story in Hindi, "Hatya" (murder), the persona of the writer lives with her family in a small two-bedroom flat in a city. She has no place to sit and write. She cannot switch on the bedroom light as it would disturb her husband. The other room is her in-laws' bedroom. She goes to the kitchen and switches on the light. And lo, at that very moment, a rat, scared by the light, runs helter skelter; utensils come crashing down; members of the family rush to the spot only to discover the sheepish *bahu* of the family standing like a culprit. They scold her for disturbing everybody. The writer says, the story she wanted to write is thus murdered – *"Hatya ho gayi*!"

Space does not mean only physical or geographical space. It also means mental/psychological space. It has always been difficult for a woman to find time and space to write. That is where the need for engendering and re-gendering becomes important so that she gets a space to indulge in an intimate encounter with the self, recognition and an openness whereby she can achieve self-definition. She becomes a crucial mediator between the word and the silences, between forms of feminine 'ecriture', to use Cicoux's term, and a culture that would deny them in the most violent manner.

Now we come to the question of borders and boundaries. Borders and boundaries supplement each other – in geography and topography, in life and relationships, in the mind and the body. The moment a border is created, the territory is demarcated – this is mine, that is thine; I and you; never 'we'. "Borders," says Jasbir Jain, "are significant markers of nation formation and go on to create communities and identities. Borders mark territorial limits, define cultural practices and signify ownership and belonging" (Jain, 2009:1). In her novel, *Cracking India* (also titled *Ice-Candy Man*), Bapsi Sidhwa interrogates the very concept of dividing the country by the stroke of a pencil. *Ice-Candy Man* is a powerful reconstruct of the traumatic events that destroyed the pace of a normal individual's life. Amitav Ghosh calls it *Shadow Lines*, signifying unreality generated by political motives.

But once the line is drawn, identities are formed and ownership is asserted. What remains is the psychological damage with which the public struggles. The political realities of India's partition impinged further on the gendered realities of the affected communities when women suffered – rape, rejection and second separation. Those who have read Sadat Ali Manto's short story "Open It", shiver at the pain and torture of the sufferers of partition. The significance of such incidents lies in the way in which the author shifts the concept of identity into a different spatial mode, suggesting that because of the created difference, on the basis of religion and culture, the barriers become permeable and crossing the border both metaphorically and literally, an act of triumph for the avenging community and an act of punishment for the victim community. In determining relationships between people, literature suggests an aesthetic in which transgression, disorientation and the uncanny become alternative strategies of living between spaces and identities.

Coming to the question of women's identity formation and the border marked for her, we realise that it signifies limits set for her – a kind of *Lakshman-rekha*. Thus, metaphorically, boundaries suggest moral distinctions between what is permissible and what is possible; what is acceptable and what is unacceptable. Restrictions and restraints draw the identity line and create the concept of the 'other'. Identifying someone as the 'other' means marginalising him/her vis-à-vis the 'self' that is 'me' and is tantamount to creating distance and allotting 'spaces'. Space can have spiritual as well as psychological connotation; likewise, it can be social and mental. To feminist theorists, relationship between gender and space is significant. When first confronted with literature on the nature of space, one is puzzled by obvious alternatives: Real and perceived; phenomenal and behavioural, ideal and material. These are broad and contrasting conceptions and within them, there can be other related sets denoting place, religion, gender, location, locale and situation.

In literature, representation plays a significant role in engendering and re-gendering women. The dictionary meaning of representation is: The act of representing or the state of being; something that represents as an image or a symbol like a verbal or pictorial portrait.

Imagination and preconceived notions play an important role in the creation of representation. Representation deals with the questions: Who can speak, on behalf of whom, for and about whom? In feminist studies, this debate has been going on for a long time. For feminists, one of the important aspects of representation is to enable women to re-present themselves because they have not been able to speak for themselves or present their viewpoints; or they were relegated to the margins.

Representations are based on social reality and hence are social constructs. That is to say, there are always taken-for-granted assumptions and beliefs behind every representation. Let us put it in the form of a question: What do we visualise when we utter the word 'woman'? The answer to this generates a stereotypical image: A person bound to domestic work, one who nurtures and brings up children, small in size, low in IQ, home-bound, subordinate and submissive, soft and tender, needing protection and so on. Indeed, women are not represented as independent agents or individuals but as part of a context, that is, in the perspective of family, friends, and colleagues.

Women are objects rather than subjects, passive rather than active. Often, their passivity extends to victim-hood. Men, on the other hand, are represented as decision-makers – powerful and in control of things. We all know the phrase 'wear the pants/trousers'? Well, a person who 'wears the pants' controls things and makes decisions. In family and also in society, it is the male who has the prerogative to make decisions and is in a subject position. The constitution of the secondary position of women is thus determined by gender.

An interesting episode recorded by an interviewer, who was seeking a male viewpoint for his study on feminism in Africa, needs be referred to here. A 26-year-old married man, Aon, asserted during the interview that it is the man who is unquestionably the head of the house and the wife must live according to his rules. Aon was a tailor (the business not so stable) and claimed that everything in his home was his property. He, therefore, controlled what the wife and the children had too. The interviewer says that he was surprised when Aon said bluntly: "Personally, I beat my wife very often – whenever she does not obey my orders. Women, at times, behave like children and have to be 'straightened up'. They, at times, talk too much and have to be shut up!"

If you are able to represent yourself as you are, you assume the subject position and concomitantly, the rest are the 'others'. The process of 'othering' is inseparably linked to self-representation. In the patriarchal system, man is in speaking position, so he is the 'self'; all those about whom or for whom he speaks are the 'other'. The speaker has the active role and an element of control. Women have so far been deprived of this central/active role and are mute.

Feminist critics have emphasised analysing literature, media and other fields to study women's representation by both men and women. Some of the prominent texts of the 1960s and 1970s like Kate Millet's *Sexual Politics*, Elaine Showalter's *A Literature of Their Own: British Women From Bronte to Lessing*, and many others, have questioned the social frameworks in which women have been relegated to stereotypical images of passive, masochistic and male-identified. In short, they have interrogated the unquestioned acceptance of men's portrayals of women. Acceptance means submission to the existing socio-cultural structures; it also means acquiescence to the culturally constructed, one-sided view of women's reality.

So, they dug out women's tradition of writing, their history and the use of language to ascertain their need to express themselves as they are, what they are, and what they feel. That is to say, they wanted a 'voice'. Caroline Ramazanoglu, a feminist critic, opines that the most obvious principle of social transformation is to take women's own account of their experience as part of their situation thus allowing previously silent voices to be heard and also making the other side (patriarchy in this case) aware of their joys and sorrows, problems and strengths. In gender studies, the issue of representation is a significant one for research as well as for political activity. Representation covers the fields of literature, politics, sociology, anthropology, cultural studies and, in fact, many other disciplines because representation/voice is an important factor for access to power.

Gender is the key issue when discussing representation. Gender as a social category is important to explore what counts as a woman. But gender needs to be understood as something that is expressed, singled out, and positioned within theories of embodiment, their meanings and relationships. Gender involves three elements: Meaning and signification (social experience); social relations of men and women (psychological aspect); and social identity (symbolic element). These three aspects are important to understand what being a woman represents and how women represent themselves.

An important question feminists ask is, how do women represent themselves? First, being a woman represents being weak or being 'different' from the male sex. In portrayal of women, masculinity sets the boundaries. Feminists resented this and wanted to explore what being a woman represents for a woman. Representation is women's attempt to think about themselves as women. In literature, it stands for their own writings in which they can express their innermost feelings and experiences. Let us take, for instance, the experience of motherhood; it is special to women and only they can understand its joys and pains and only they can represent the experience authentically.

But representation must take into consideration its opposite, i.e. misrepresentation, which is beset with dangers. No act of representation can ever take place entirely outside of ideology. And women, as part of the existing ideology, also participate, knowingly or unknowingly, in acts of misrepresentation. For example, there was a time when woman writers were looked down upon. A woman entering the literary field was seen as a freak, someone who belonged to the margin, a second rate person, and if at all she was allowed some space, it was a special favour. When Charlotte Bronte gave her novel, *Jane Eyre*, to Robert Southey for perusal, his acerbic reaction was, "It is not women's business to write." In her novel, *The Binding Vine*, Shashi Deshpande records how a renowned poet, Venu, reacts when the young protagonist, Mira, shows him her verses. He rebuffs her saying that a young and beautiful girl should leave poetry to men;

women's poetry is to give birth to children, implying, thereby, that the patriarchal system thought of women as body and not as brain.

I am tempted to quote here what Nabneeta Dev Sen recounts in her opening lecture at a Sahitya Akademi Conference, 'Women Writing in India at the Turn of the Century'. Male writers are usually not very respectful of woman writers and their writings. Example:

> "Hi Nabneeta, I liked that earthquake poem of yours in *Desh* this week."
> "Ah, did you? Thank you! I liked it too. It is not mine though, Bijoya wrote that one."
> "Ha, ha, is that so? I thought it was you." (Sen 2009: 5-6)

In another conference, she was identified as Kanu Mishra and in yet another, as Kabita Sinha. In one case, Sen got so irritated that she decided to make fun of their negligent attitude and going on the stage, she said, "I am not Kabita Sinha, I am Mahasweta Devi." Those who knew her and understood the joke saved the situation from blowing up. The point is, as a poet says, "They read us/casually, as one reads/ the torn pages of a child's notebook, before it is made into paper cones for *Chanajorgaram.*"

So we put it thus: Representation means presenting the perspectives of those who have not been able to represent themselves. We need to understand the individual experiences of women but we also need commonalities to be ascertained and experiences to be made collective. In this context, we need to examine the ways in which representation of gender helps structure cultural perception of women. In the Indian socio-cultural psyche, a woman is deified as a *devi* or goddess. But this is also misrepresentation of a social reality because by putting a woman on a pedestal, man denies her human weaknesses, desires and wants. She is seen, not as a human individual, but as a symbol. Indian feminists like Yashodhara Bagchi, Jasbir Jain, Sussie Tharu and many others are trying to explore how Indian women define themselves and are defined as women within various systems of representation. Allowing women to define themselves means accepting their identity and identity is important to understand them as unique individuals in their own right.

When women attempt to redefine their roles and expand their arena of choices, patriarchy resents these notions because their actions may necessitate re-structuring of the existing social and political norms and also because redefinition may lead to self-definition and identity formation. Articulation of identity has to be understood as the function of historical, social and material circumstances.

Women try to capture in their writings those moments that are crucial in the shaping of their identity as gendered subjects. Notions of femininity, of respectability and of honour often shape the reaction of women to questions of sexuality. In the film, *Fire,* for example, Deepa Mehta uses feminism in a manner that does not fit into the normal, socially accepted discourses. The film generated feminist discussion regarding the representation of lesbianism and also incurred the ire of society. Representation in cinema requires both language and visual depiction; other disciplines like literature and social sciences depend on language to delineate reality.

Language is not neutral; it is not simply a carrier of ideas but is a shaper of ideas. It is a means of self-expression and is co-related with socio-cultural facts of one's upbringing, environment, class, nationality and other basic attributes. This is called 'language socialisation' which means people learn how they are expected to categorise themselves socially and how people in those categories are expected to talk. Thus, women, as a category, employ language differently from the way men do and hence, 'women's language' is believed to display their female identity. Researchers claim that language and gender are related to the association between language and the social contexts because it is the social context that forms our concepts of how men and women use language. Feminist linguist Deborah Cameron argues that the activities that women and men participate in and the way they communicate provide clear indication of the relationship of language to gender. Men and women differ in their communicative manner and critics assert that language ideology plays an important role in the gendered use of language.

Language is the main reason on which all claims of gender difference depend. Some linguists believe that women are aware of their low status in society. So, women use language that is powerless and has been developed as a way of surviving without control over economic, physical or social reality. Critics also claim that women speak politely because they are secondary in status to men and because politeness is expected from the inferior towards the superior. However, we cannot make sweeping and generalised statements in this matter as language use is governed by social and cultural factors and is different in different cultures. But it cannot be denied either that women are at a disadvantage because language is a male construct and women have to use the 'master's language' as it is; they do not and cannot have their own language.

In gender studies, discourse analysis answers questions about social relations, personal and social identity, self-definitions, dominance, and social and personal adaptation. Let us take an example to see how discourse analysis works. In the novel titled, *Forever Free*, written by Raji Narasimhan, some woman characters talk about the protagonist thus, "It's like a man, she is, arguing and going to the bottom of things. She's like a man." Even if we do not know the context, we can analyse these two sentences to reveal different aspects: (1) The protagonist is a woman ('she' pronoun); (2) she appears to be a strong character (personal trait); (3) women do not argue as vehemently as men do (social expectation); (4) women do not go to the bottom of things (generalised ideology); and (5) the women who are discussing the absent woman seem to resent her because she is 'like a man' or maybe, they appreciate her for her independence. The above two sentences from the novel are silent on the nuances we brought to light by our analysis. This means that 'silence' also articulates many things and needs to be understood.

'Silences' mean those statements that are in the background, that are not expressed but have been understood. In the above example, the speakers did not say that they are jealous or that they disapprove of the protagonist. They are silent on the issue but we can notice it. Silence also means lack of voice, overt restriction placed on speaking or prohibition imposed on expression, exclusion from highly valued forms like rhetoric, or public speaking. Terms such as 'absent' women, 'missing' women and 'invisible' women indicate women's silences. The idea behind discourse analysis is to detect silences. In her book, *Discourse Analysis*, Barbara Johnstone observes, "learning to notice silences means learning to 'de-familiarise' the familiar." (Johnstone 2002: 58-61). This is one way that may lead to giving power and control to the 'silent other'.

Cinema, literary works, art, rhetoric and many other fields offer a lot of potential to study and understand the role silences play. For example, in the film, *Peepli (Live)*, discourse analysis can show us many 'unsaid' things about our present socio-political system. Another example can be the 1982 film *Arth* where silences and social insecurities of a single woman and the psychological problems generated by guilt can be analysed. Silences can be read in male writings as well. In Manohar Malgonkar's novel, *The Princes*, the king often boosts the morale of his son by exhorting him to 'Be a man'. It means, 'be strong and brave'. The sentence does not say 'women are weak'; it is silent but reveals the intended meaning and the social expectation that men should be strong.

In re-gendering, women's oral traditions, storytelling and myth-making abilities need to be studied. One of the challenging fields is revisionist myth-making. Revisionist myth-making redefines the culture and women's place in the culture. They speak of the past, re-interpret it and re-write it from the woman's point of view. If the woman was absent from a mythological tale, revision gives her the desired presence. Feminist critics consider it an act of survival.

Issues raised by woman writers of the sub-continent vary from gender-bias to violence; exposure and censure of social evils like poverty to the economic and moral iniquities to the inhuman caste bias and dehumanising superstition; from corruption to nepotism, as also from nostalgic reminiscences to the alienating forces of modernism. Kishwar Naheed, a Pakistani poetess, vociferously defends those women who have the guts to speak up. Society may call them 'sinful women' but they, though unconventional, have the courage to articulate and stand for truth.

Another poetess, Razia Hussain from Bangladesh, interweaves the horrors of war with the promise of life. Razia Hussain, a Dhaka-born poetess, is a peace activist and feminist. In "The Sound of Leaves", she depicts war and the resultant violence. Her motherland is ravaged by the hand of the power-hungry male but the earth cannot be left barren. Neither the poet, nor the earth, which is being torn apart in the battle to possess her, submits passively to violence:

> War, only war all twelve
> months of the year. Our eyes are cactus plants,
> the thorns growing inward to pierce our tenderest nerve.
> Still, sometimes the sound of leaves makes me open my eyes to the sky.
> Again the mind begins to build its nest among quiet wings,
> the shadow of the Shal tree falls green
> over my house over the smell of this warm, wet earth.

In the field of fiction, woman writers like Bapsi Sidhwa, Taslima Nasrin, Yasmin Gooneratne, and, of course, woman writers of India, have made remarkable contribution. In *Ice-Candy Man*, Sidhwa lets her own nightmare be expressed through Lenny: "I feel no pain, only an abysmal sense of loss – and a chilling horror that no one is concerned by what's happening". This 'no one' is none but the political leaders responsible for the partition. Now, more than six decades after Independence, things have not changed much. There is still communal disharmony, religious fanaticism and political connivance. Many woman writers are focusing on issues such as these and countering the charge that women's writing is only sentimental crap.

In the postmodern scenario, women are taking to what postmodern critics call 'dedoxification' or reverse writing. They are writing from the angle of the subaltern – women, tribals, servants, dalits and others. The fact that woman writers are speaking up and breaking age-old traditions reveals the beginning of 'the freedom of the mind'. According to Virginia Woolf, it denotes the possibility that, in course of time, the mind will be free to write what it likes. Self-assertion of authors or their protagonists' consciousness of their plight does not necessarily proclaim feminism. But the sporadic portrayals of resistance and protest derive from the deeply felt lack of expansion and the necessity to breathe. These are, indeed, brave women. By thus writing resistance, woman writers are opposing evils and empowering themselves.

Writing means breaking the silence, speaking up. And speaking up is a difficult proposition for women even in the present day scenario. Society is not yet ready to hear other voices or exonerate the woman. But then, patriarchy is not to be judged by a few bad men who beat their wives and a few good men who care for them. It is to be seen in its totality as working within the politics of power structures. It is this power structure that woman writers take cognizance of when they write texts showing options open to women. In depicting women's representation, literature is expressing a plea for control, empowerment and women's agency. Women must, therefore write, as Helene Cicoux says. Women must write so that they are read by men. I am tempted to quote what Nabneeta Dev Sen says:

> We need to be read and known by our fathers, by our husbands, our lovers, our sons. If we want to change the system we must work together as family units. It's a period of transition that we are passing through and we must carry our families with us. Our life work must not look like a proselytisation plan, the change should come within, as a natural process of development in the family structure. We shall always remain woman writers; no amount of backlash can put us back into our shell again. It is broken. Once for all, we know who we are. We want to know what the whole of life is all about. Not only ours, but theirs as well. (Sen 2009: 18)

Works Cited

Jain, Jasbir. Crossing Borders: Post 1980 Subcontinental Writing in English. Jaipur: Ratwat, 2009. Print.

Johnstone, Barbara. *Discourse Analysis*. UK: Blackwell Publishers, 2002. Print.

Ramazanoglu, Caroline, and Janet Holland. *Feminist methodology: Challenges and choices*. Sage, 2002. Print.

Sen, Nabneeta Dev. "Women Writing in India at the Turn of the Century." *Growing Up as a Woman Writer*. Ed. Jasbir Jain. New Delhi: Sahitya Akademi, 2007. Print.

CHAPTER TWO

APHRA BEHN AND COLONIAL SPACE

MARGARETE RUBIK

Aphra Behn was an English poet, dramatist and novelist at the end of the seventeenth century. She was the first woman writer to make a living from her pen – a career which, before, had only been open to men. She was forgotten for two centuries because of changing tastes and moral precepts until she was put on the literary map again by Virginia Woolf. Woolf exhorted all women to "to let flowers fall upon the tomb of Aphra Behn… for it was she who earned them the right to speak their minds" (Woolf 59).

In her life, Behn transgressed borders in more than one sense. In the seventeenth century, poetry was an 'acceptable' genre because women could compose it at home and circle it among their acquaintances in a non-competitive manner. But by writing for the public stage, Behn entered a space traditionally connoted as male. At that time, Gallagher argues in "Who was that Masked Woman?", a female playwright writing for the public theatre was equated with a public woman, that is, a whore. When it came to women, Restoration audiences and critics tended to conflate the public and private space. There was a prurient interest in the private lives and affairs of actresses and female playwrights while no one cared for the behaviour of men. It was assumed that the plays they wrote and the roles they starred in reflected their private lives and morals. A woman who, like Behn, wrote sex comedies, was assumed to have experience in the field. The fictional space of the theatre was seen as a direct mirror of her lifestyle. The private sphere of the female dramatist and the fictional text worlds which she invented were fused and became inseparable. Despite such wide-spread prejudices, Behn managed to cross class borders by being accepted in the royalist aristocratic circles.

But even before she began her career in the theatre, Behn transgressed gender borders: She was employed as a government spy in the Netherlands and ended up in debtors' prison because the king did not pay her. Earlier still, in her youth, she travelled to Surinam, an English colony at the mouth of the river Oroonoko in South America.[1] She described this experience in her novel, *Oroonoko*, a fictional tale with some semi-autobiographical elements.

It is characteristic of the gendering of spaces even in the 20th century that several critics accused her of being a liar, of never travelling across the Atlantic at all and of having copied her descriptions from contemporary travel reports.[2] Of course, it was unusual for women to travel such distances. Female overseas travel only became popular in the second half of the 19th century and even then, the adventures of single woman travellers like Isabella Bird and Mary Kingsley were curiosities. If we look at the18th century, female travellers like Lady Mary Wortley Montague or Anna Maria Falconbridge accompanied their husbands on official journeys. We do not know in what company Behn travelled to Surinam but by now, it is fairly certain that she did go there – because of her accurate knowledge of political functionaries and the fauna and flora there as well as some expressions from indigenous languages.[3]
The novel, *Oroonoko*, is a fictional tale about an African prince lured on board a slave ship and transported to Surinam where he is united with his lost love and later tortured to death after a failed slave rising. The narrator, who bears some similarities to the author, only occupies a role in the background. The story consists of two parts connected by the two places of action: Coramantien in Africa, in what is today Ghana, and Surinam in South America.

From the formalised descriptions, it is fairly clear that Behn was never in Africa herself. She may have gained knowledge of that part of the world either through travel reports or, more probably, through the clichéd conventions of Orientalist plays or, perhaps, through the narration of the African prince Oroonoko – if indeed, such a person ever existed. In the novel, this warrior prince initially operates in the outdoor space, committing heroic acts in battle. But the story focuses on the indoor space as soon as his beloved, Imoinda, is forced to enter the old King's harem. The harem was already a fascinating place to writers of the Restoration period and became increasingly so after 1706, when the first English translation of the *Arabian Nights* was published. Western tales about the harem often follow a stereotyped pattern, featuring a sexually aggressive sultan, an innocent virgin and a true lover who tries to free her. In Behn's novel, Oroonoko gains entrance into the forbidden space of the harem and

---

[1] For Behn's life, cf. Maureen Duffy's biography.
[2] Cf. for instance Bernbaum.
[3] Cf. Dhuicq 40-41.

manages to consummate his marriage with his betrothed. But she is sold as a slave in retaliation. The lovers are then re-united in America.

The second part of the narrative is set in Surinam. The narrator passes herself off as an eye-witness, fusing tragedy, romance and adventure narrative with ethnography and natural history. She not only claims to have been Oroonoko's confidante and hence his appropriate biographer but also styles herself into an explorer of indigenous South American tribes, goes on tiger hunts and is the only survivor of the colony's turbulent history to tell the hero's story. Thus, she enters a sphere to which, at her time, only privileged males had access.

In her description of the fauna and flora of Surinam, the narrator praises the abundance and beauty of the country. The colony seems to her a treasure-house, full of marvels and useful produce from spices, and fruits to exotic animals and valuable timbers.[4] As is typical of colonial fiction,[5] aesthetic pleasure at the beauty of the land is fused with a calculation of the economic profit its exploitation could yield to the English – which explains her frustration at the fact that the English lost the colony to the Dutch a little after her departure from Surinam.

As in so many travelogues, Behn waxes enthusiastic about the wonders of the New World – the "wonderful and surprising forms, shapes, and colours," (Behn 75) – exotic sights and smells, and trees bearing blossoms and fruits at the same time. In the course of the narrative, however, the idyllic atmosphere gives way to shocking details of mutilation and torture and the horrible smell of rotting bodies. If Behn started out by painting Surinam as a place of exotic beauty, she ends up by depicting the colony in the hands of a bloodthirsty lower class mob devoid of any humanity.

Behn constantly juxtaposes dichotomous images of the colonial space even in her description of colonial relations with the South American Indians. At first, Surinam seems to be a contact zone of mutual love and respect. The British colonists are said to "caress [the Caribs] with all the brotherly and friendly Affection in the World" and to live with them in "perfect Amity" (Behn 75). The true reason for this profession of friendship, however, is economic expedience and fear that the natives, by far outnumbering the white colonists, might rise in rebellion and overrun the settlers (as, indeed, they later did under Dutch rule). Since the colonists do not dare to force the indigenous population into slave labour, they import Africans to work on the plantations but dread the possibility of a revolt on their part as well. Far from being a place of peaceful coexistence,

---

[4] Cf. Rubik "Teaching *Oroonoko* in the Travel-Literature Course."

[5] Cf. Spurr.

the colony seems rife with fear and dishonesty, exploitation and enslavement.

Similarly, contradictory impressions are conveyed by the imagery with which the Caribs are described: On the one hand their naiveté makes them seem like Adam and Eve before the Fall, giving the narrator "an absolute *Idea* of the first State of Innocence, before Man knew how to sin" (Behn 77). Pre-colonial Surinam is thus constructed as a paradisiacal space, contrasted to the corruption of white civilization. Yet, the narrator, nonetheless, seems to prefer the 'fallen' civilization of her own sophisticated culture to this stone-age innocence. On the other hand, the self-mutilation of the Indian war-captains, who slash away at their ears and noses to prove their prowess, jars with the image of Eden as does the news of their later cruelty during an insurrection against the less diplomatic Dutch colonists, when "they cut in pieces all they could take… hanging up the mother and all her children about her; and cut a footman … all in joints, and nailed him to trees" (Behn 120).

Though Behn refuses to sentimentalise the Caribs, it is important to note that she never uses terms like 'savage' or 'primitive' for them. The phrase "*below the wildest Salvages* [i.e. savages]" (Behn 126), which Oroonoko employs in his stirring address to the other slaves, is reserved for the whites and their shocking cruelty and treachery. It is they who introduce the Indians to the concept of lying; they betray Oroonoko more than once and torture and kill him in a bestial way. In a biblical scenario, they are the true embodiments of evil. However, even the noble savage, Oroonoko, commits a shocking cruelty when he decapitates his pregnant wife in order to save her from falling into the hands of the pursuers. The colony of Surinam, then, is both paradise and hell.

The description of the indigenous American population is also interesting for another reason. The narrator and some of her friends (including Oroonoko, who is, in this scene, subsumed within the white travel party) make an excursion to an Indian village which has supposedly never been visited by white people before. Instead of assuming the typical attitude of explorers and travellers directing their curious gaze on foreign ethnicity and thereby asserting their superior subject position vis-à-vis the foreign objects of scientific investigation, Behn reverses the power relations. Her narrator is willing to surprise the American Indians with "something they never had seen, (that is, White People)" (Behn 121), thus turning her white travel party into curiosities for the amusement of the indigenous villagers. To be sure, the white tourists come for exotic 'Diversion' (Behn 121), but they are also aware of being, in their gaudy robes, a sight of 'numberless wonders' (Behn 121) for the Indians. For once, the European travellers are keenly aware of being somebody else's 'other' and of being subjected to an objectifying gaze. Indeed, the Indians wonder whether "those things [the Europeans] can speak", have "Sense, and Wit" and can talk about "of affairs of Life" (Behn 122).

By making this trip into a remote Indian village and entering into close physical contact with foreign ethnicity, Behn's narrator, once more, claims a space that is not normally open to women. The house is the place generally allotted to women. For all her social pride at occupying the best house in the colony, she never sees the domestic space as a sanctuary. Rather, it is a site of deceit where she spies on Oroonoko and tries to keep him amused so that he will not think about a rebellion. It is also a space where she feels most vulnerable to attack. The woman who ventured into the jungle to go on expeditions and tiger hunts now fears that, if she stays in her house, the revolting slaves will cut her throat – in spite of her supposed friendship with their leader. Although she firmly placed herself in a male position before, she suddenly uses the female privilege of timidity and flees down the river when the revolt breaks out, removing herself from the place of action. Later, she bitterly regrets her desertion since she claims that, by her authority, she could have saved the hero's life. But she later re-assumes the masculine role by acting as Oroonoko's biographer. Behn's text consistently deconstructs stereotypes of race and gender.

Let us, in comparison, briefly consider some 18th century female travellers: Lady Mary Montague is best known for her description of a Turkish harem in *Turkish Embassy Letters*. Unlike European men, she could enter the private space of Turkish women and was invited to participate in their pastimes. Thus, she visited a women's bath (although she herself refused to take off her corset). She also learned to appreciate the veil as a protection for women who wanted to walk out into the streets unknown, clad in garments which screened them from the public space in which they moved. Behn, we have seen, craved access to spaces reserved for men. Montague, in contrast, valued the all-feminine sphere about which prurient men could merely speculate.

Anna Maria Falconbridge, at the end of the 18th century, accompanied her husband to Sierra Leone to defend and protect the rights of the Nova Scotia black loyalists who had been invited to settle there. She published a collection of letters about her experience. Unlike her Restoration predecessor, her impression of the tropics is mainly negative. The country is not Paradise but an 'inhospitable' wilderness in which she experiences terror and "hardships unprecedented" (Falconbridge 81). Like Behn's narrator, Falconbridge, too, travels into the interior of the country – only to be terrified by a hostile tribe which threatens to kill the white party if it does not leave immediately. She is equally scared of the local African ruler and during an invitation to his feast, is reduced to gibbering hysterics. On the other hand, she refuses to be confined to a narrow cabin on board the ship, which she likens to a prison. In fact, Falconbridge feels safest in the comfortable houses of the factors – no matter that they are slave traders. Although her vision of the colonial space is much more negative than Behn's, she, too, blurs the borders between the colonial space and the metropolis by presenting the managers of the Sierra Leone Company as much more hard-hearted, stubborn and cruel than the indigenous kings.

We have to wait a long time before we meet another female traveller as open to encounters in the colonial contact zone as Aphra Behn. The case of the Austrian Ida Pfeiffer comes to mind. She, in the middle of the 19th century, alone and fearlessly travelled to the head-hunters' territory in Sumatra. Pfeiffer's arrogation of the male privilege of scientific travel to remote places of the earth and her insight that the colonial space offers an opportunity for women to transgress narrow gender roles and to escape confinement in the domestic sphere is indeed reminiscent of her well-known Restoration predecessor.

Works Cited

Behn, Aphra. *Oroonoko, The Rover and Other Works*. Ed. Janet Todd. London: Penguin, 1992. 73-141. Print.

Bernbaum, Ernest. "Mrs Behn's Biography a Fiction." *PMLA*. 28.3 (1913): 432-53. Print.

Dhuicq, Bernard. "New Evidence of Aphra Behn's Stay in Surinam." *Notes and Queries* 4.1 (1995): 40-41. Print.

Duffy, Maureen. *The Passionate Shepherdess. Aphra Behn 1640 – 89*. London: Cape, 1977. Print.

Falconbridge, Anna Maria. "Two Voyages to Sierra Leone (1794)". *Maiden Voyages and Infant Colonies*. Ed. Deirde Coleman. London: Leicester UP, 1999. 45-168. Print.

Gallagher, Catherine. "Who was that Masked Woman? The Prostitute and the Playwright in the Comedies of Aphra Behn," *Rereading Aphra Behn. History, Theory and Criticism*. Ed. Heidi Hutner. Charlottesville: UP of Virginia, 1993. 65-85. Print.

Rubik, Margarete. "Adventurer, Western Observer and Female Scientist: The Conflicting Interests of Ida Pfeiffer's Travelogues. "*Stories of Empire. Narrative Strategies for the Legitimation of an Imperial World*

Order. Ed.Christa Knellwolf and Margarete Rubik. Trier: wvt, 2009. 137-55. Print.

—. "Teaching *Oroonoko* in the Travel-Literature Course." *Approaches to Teaching Oroonoko,* Eds. Cynthia D. Richards and Mary Ann O'Donnell. (Modern Language Association) Forthcoming 2013. Print.

Spurr, David. *The Rhetoric of Empire: Colonial Discourse in Journalism, Travel Writing, and Imperial Administration.* Durham: Duke UP, 1993. Print.

Woolf, Virginia. "A Room of One's Own." [1929] *A Room of One's Own and Three Guineas*. London: Chatto and Windus, 1984. Print.

CHAPTER THREE

RE-GENDERING THE HARDBOILED GENRE: THE CRIMINAL FEMME FATALE IN RAYMOND CHANDLER'S FICTION

MAYSAA JABER

Raymond Chandler is a canonical writer of hardboiled crime fiction, a distinctively American genre of crime writing that developed in the interwar period of the 1920s and 1930s, fully maturing in the 1940s. The genre depicts a world of crime and corruption usually with a male detective oscillating between the rich elite and the seamy underworld of gangsters and dangerous women. Chandler's work introduces a large number of what I term 'criminal femme fatales' – women who go beyond the arena of dangerous sexuality to enter the realm of criminality. These criminal femme fatales use their sexual appeal and irresistible wiles both to manipulate men and to commit criminal acts, usually murder, in order to advance their goals with deliberate intent and full culpability. This essay aims to complicate the notion of the femme fatale as simply a seductive and threatening woman by arguing that there is more to the femme fatale in American hardboiled crime fiction than is usually allowed for. While feminist scholarship regards the femme fatale as a sexist construction of male fantasy and treats her as an expression of misogyny that ultimately serves to reaffirm male authority, I aim to open the closed doors behind which this archetype of the femme fatale is trapped.[1] I reconsider her role,

[1] While the femme fatale has been carefully studied in cinema, she has not been the subject of sustained critical attention in crime fiction. It is indeed film scholarship that has played the most vital role in formulating critical attitudes regarding the femme fatale. Feminist film scholarship, which flourished in the 1970s and 1980s, saw the classic film noir of the 1940s and 1950s as ripe for analysis, reading the femme fatale figure of this film genre from a range of perspectives, with many analyses circling back to the question of the femme fatale as an objectification of male desire. Laura Mulvey's theorisation of the male gaze is perhaps the most

beyond her image of an 'evil' seductress, as a complex character, a force that drives the narratives and a figure that prepares the ground for more nuanced readings of the constructions of gender and sexuality in the genre.

Specifically, I will examine representations of this figure that open a space for imagining female agency. By agency I refer to the ways that criminal femme fatales demonstrate power, intelligence, and independence as they successfully mobilise their skills in a male-dominated milieu (both textually and at large). As such, these female characters break normative conceptions and expectations of gender roles by challenging the pattern of female submission, domesticity, and dependence, and also break the fixity of the genre that is often described as 'masculine'.[2] My argument, therefore, offers possibilities for examining the femme fatale against the grain, expanding the margins and loosening the tensions between gender and genre, and inviting a more reparative feminist reading of the role of women in hardboiled crime fiction beyond the discourses of objectification and misogyny.

Chandler consistently depicts murderous women whose representations are, nonetheless, interconnected with his detective, Philip Marlowe, who occupies a central position in Chandler's fiction[3]. I will argue that while Chandler's criminal femme fatales do serve as foils to Marlowe in his professional and moral capacity, they do not do so simply to underline the detective's power and 'heroic' qualities. Rather, the criminal femme fatale destabilises the status of the detective as 'hero' and questions his mastery. The narratives achieve this through the representational dynamics that Chandler uses to portray his female characters. They are not always visible in the sense that they do not dominate the narratives with an overpowering presence and we do not see them commit criminal acts. Their agency is located, however, in what I will call (in)visible roles. That is, despite or because of indirect avenues through which female agency is mobilised (whether through absence, or retrospective reading of agency, which I will

---

influential and oft-cited example used in feminist scholarship on the femme fatale. For studies of the femme fatale in film noir see, for example, Kaplan, Maxfield, Hanson and Grossman.

[2] Hardboiled crime fiction is considered "male oriented" both as far as its notoriety as a "masculine" genre and the history of its literary criticism are concerned. Masculine toughness and masculine codes are essential in hardboiled narratives, and are part, as Jopi Nyman argues, of a larger ideology of masculinity: "the focus is now on the different constructions of gender and their relationship to culture and history rather than on an individual writer's intentions" (39).

[3] All of Chandler's novels except one – his last finished novel *Playback* (1958) – contain criminal women, although *The High Window* (1942) and *The Little Sister* (1949) do not present their female protagonists as femme fatales.

explain in this essay), the narratives invite us to infer the criminal femme fatale's empowerment through her interactions with Marlowe which ultimately prove his vulnerability to her unchecked power.[4] Chandler also plays with the reader's expectations of the role of women as criminals; his narratives interrogate the ways in which female characters appear, or, more accurately, do *not* appear, and negotiate their agency as criminal femme fatales.

Marlowe is described by both his creator and some critics using superlative adjectives that endow him with 'knightly' and 'heroic' qualities.[5] Chandler describes his creation as "a man of honour… the best man in his world and a good enough man for any world" (The Simple Art 14). Marlowe does not succeed in rescuing the 'heroine', however, primarily because the women in the stories do not need to be rescued. Chandler's mission statement for detective fiction is a search for 'hidden truth', (The Simple Art 15). Yet his detective fails to solve the mystery or restore order; Chandler's fictional world thus remains highly fragmentary. As such, I problematise the reading of Marlowe as a 'hero', and suggest that the detective's missions always face challenges posed by criminal women that ultimately get him out of control and, arguably, questioning his mastery. I also argue that Marlowe's detective shield and his morality as a "man of ideals that, although antiquated, continues to rule his behaviour" (Abbott, *The Street was Mine* 48), are merely masks behind which he hides his weakness for women in particular. The anxiety that Marlowe displays towards women is not only sexual, but also relates to power. The detective, facing constant threats including death, in Lee Horsley's words, "undergoes trials which in fact subvert the stereotypical

---

[4] To the contrary to my argument, there are a number of critics who advocate readings of Chandler's narratives as assertive of the detective's masculinity over the femme fatale. For example Megan Abbott states that "hard-boiled masculinity relies on the defeat of the femme fatale for its own existence, Marlowe's masculinity is structured rather as a house of cards, precariously built through a ritual transformation of the 'feminine' into a contagion to be avoided" ("Screening Marlowe's Masculinity" 307). My reading of the role of the femme fatale differs from Abbot's view that Marlowe seeks to "eradicate or at least contain these lethal women."

[5] There is a massive amount of critical scholarship about Philip Marlowe as a "knight" or a "hero." For example, Philip Durham describes Chandler's hero as "passionately ethical" (86) and asserts that Chandler adheres to the characteristics of the "hero," which include "courage, physical strength, indestructibility, indifference to danger and death, a knightly attitude, celibacy, a measure of violence, and a sense of justice" (81). For similar readings on Marlowe, see Knight ("A Hard Cheerfulness" 80) Skenazy (37-38), Lid (42), and Marling (81-83).

image of male mastery… his tough image seems more like a reflex of fear" (73). Marlowe's vulnerability and his fear of losing authority function as a mirror that reflects and reinforces women's agency.

This essay, therefore, addresses the (in)visibility of women's agency in relation to Marlowe through a close reading of Chandler's fourth book and probably the least-studied one, *The Lady in the Lake* (1943). It is a story that focuses on the disappearance of women – women who are neither clearly present nor vividly visible. The book begins with wealthy businessman Derace Kingsley asking Marlowe to find his estranged wife, Crystal, who has sent a telegram declaring that she plans to marry Chris Lavery, her gigolo boyfriend. As Marlowe works the case, he becomes involved with the disappearance of another woman, Muriel Chess, and Chandler thus delineates a parallel plot based on a second absent woman. Marlowe discovers a body, believed to be Muriel's, in Lake Fawn but it is only at the end of the novel that we learn that the lady in the lake is, in fact Crystal, who was killed by Muriel. We also learn that Muriel's real name is Mildred and that she has changed her name and identity several times without Marlowe or the reader realising that people who appear to be different characters are, in fact, one woman.[6] Mildred thus turns out to be the criminal femme fatale of the book, a dangerous murderess, who is finally killed by her cop ex-husband, Al Degarmo, who is himself killed by the police in the novel's climactic scene.[7]

What distinguishes *The Lady in the Lake* from Chandler's other novels is its narrative complexity, established through the use of two parallel settings – Bay City and Lake Fawn – and the interplay of identities between the two women. Chandler hence sketches the criminal femme fatale through the dichotomy of absence/presence and false identities. Mildred's evasion of justice by changing her name and moving to a different place, and the fact that we do not see her as Mildred in the book, endow her with the capacity to act freely (and criminally) without facing any consequences. Her agency is, therefore, not direct in the sense that the

---

[6] Given the confusion around identities and names of the women in the book, I will use "Mildred," instead of Muriel (one of the fake names she assumes) unless in quotes or otherwise needed.

[7] Tom Hiney (88) shows how Chandler based *The Lady in the Lake* on a real crime that took place in Santa Monica involving a doctor named George Daley, who faced a murder trial five years after his wife allegedly committed suicide. Chandler takes the details from a front page story that claims that Daley drugged his wife and placed her in his car with the engine running to make it appear that she died from inhaling exhaust fumes. Daley, however, was acquitted as no evidence was found to convict him.

reader cannot identify it immediately. Rather, it may only be identified retrospectively, in a second reading of the book or with the realisation, at the end of the narrative, that it was Mildred who committed the crimes and was able to deceive everybody around her including the detective.

One device that Chandler employs to demonstrate the dynamics of the women's effective absence in relation to the detective's struggle for power is the blurring of identities. Jerry Speir suggests that 'mistaken identity' is an old trick in the genre of detective fiction but is cleverly used by Chandler so that the reader is "aghast at his own willingness to be duped" (51). Chandler, according to Speir, tries to maintain the "illusion, and even to mock what he is doing, without the reader being wiser until the very end" (51). Chandler's narratives consistently present disguised women in a "series of murderous masquerades" (Knight, *Crime Fiction* 120). This disguise not only adds duplicity to the qualities of the criminal femme fatale but also extends complexity through multiple layers of identities (and, concomitantly, possible readings).

So, in the novel's quest for an answer to the titular question – who is the lady in the lake? – Chandler appears to delineate something more akin to the classical detective story where the search for the missing person is central to the plot. However, Chandler problematises the classical formula by negating his detective's effort to assert power, which is an important element in classical detective fiction. The criminalisation of the femme fatale also works to countermand the classical formula whereby the dynamic of the detective's power turns on the woman's lack of it. Hence, the two factors that capture the criminalisation of women in *The Lady in the Lake,* besides the fact that she commits murder, are the detective's loss of power and control and the absence that, in fact, marks her agency. Instead of putting Mildred before the criminal justice system, Chandler has her killed by Degarmo, a representative of the law, though for personal reasons rather than for her crimes. The narrative, therefore, places the criminal femme fatale on an equal footing with her killer, as both end up dead, while also inviting a questioning of the efficacy of law and order, established with an image of an absent murderous woman that is set against a powerless detective and a dead policeman.

Mildred is not the only woman who is absent in the novel. Crystal is another dangerous woman, although not a murderess, who is positioned as a foil to Mildred. Though we do not meet Crystal, we learn of her character through the narrations of her husband and Marlowe. Marlowe surmises "that she is young, pretty, reckless and wild. That she drinks and does dangerous things when she drinks. That she is a sucker for men and might take up with a stranger who might turn out to be a crook" (*LL* 13).[8] She also enjoys financial independence: Derace Kingsley tells Marlowe that "she has her own money and plenty of it" (*LL* 11). Her fortune, which comes from a corporation with "valuable oil leases in Texas", establishes money as a corrupting force, a theme that runs through Chandler's work. What is most significant here, however, is that money is what gives women in this novel the power to do what they want. Crystal 'plays around' and Lavery, her boyfriend, is only one of the 'playmates' that Kingsley describes. In the dysfunctional relationship between Crystal and her husband, the novel delineates the disintegration of family values by positioning the woman outside of the feminine domestic sphere. She also seduces Bill Chess, sleeps with him and wrecks his marriage to Muriel.

The narrative thus paints a picture of Crystal as a villainess, but reverses this expectation when we discover that she was killed by Mildred. The confusion in identities, moreover, blurs the boundaries between victimhood and criminality, thus opening ways to see female agency in the book through the dynamics of fake identities. As the novel progresses, it becomes increasingly difficult to tell the two women apart in a physical sense: "[Crystal] is blonde like Muriel, same size, and weight, same type, almost the same colour eyes. But, brother, how different from almost then on in" (*LL* 31). The narrative intensifies this conflation by making both women threatening, which also highlights the issue of female agency in the novel. That is, Chandler makes Crystal *look like* a scheming and threatening criminal seductress from the outset, only to surprise his readers with her murder by another woman, the similarly dangerous Mildred. However, Crystal does not acquire the status of a helpless victim. It can be argued that Crystal is a femme fatale of sorts but not a criminal one – that role is played by Mildred.

Mildred not only kills Crystal but before moving to Lake Fawn, changing her name to Muriel and marrying Bill Chess, she also kills Dr Almore's wife with an overdose of drugs while she works as a nurse in the Almore home. Mildred, hence, transgresses the medical profession and also the boundaries of the traditional female role (that of care- and life-giver). Mildred merely uses her feminine role of a nurse as a convenient gateway to the masculine realm of crime while, nevertheless, maintaining her sexual appeal as a femme fatale. However, Mildred's ability to manipulate people, especially men, is closely associated with the fake identities she assumes, and ultimately permits and constitutes her

[8] All citations from *The Lady in the Lake* will be from Raymond Chandler, *The Lady in the Lake* (1959). The abbreviation *LL* appears before the page number in all quotes from the novel.

criminality. As Marlowe testifies, "If Muriel Chess impersonated Crystal Kingsley, she murdered her. That's elementary" (*LL* 183). The narrative invites us to locate Mildred's agency in her unfixed identity which gives her the liberty to act as she pleases without repercussions although the changes in her identity occur outside of the narrative, and it is only via Marlowe's narration in the final pages of the book that the 'truth' is revealed. As Carl Malmgren elaborates,

> …she is, of course, the missing link, the sole element tying together the various missing women in the novel. She, in effect, makes women both appear and disappear… The object of Marlowe's quest is, thus, always absent, *unrecoverable* until she has been converted into actual object, naked and dead in the lake (Crystal Kingsley) or naked and dead on the bed (Muriel Chess) (96, italics added).[9]

For Malmgren, Mildred does not become an 'actual object' until she is a dead body. I argue, however, that it is precisely this 'unrecoverable' absence that authorises Mildred's agency. It is also important to note that Chandler builds a story that is devoid of gangsters who are strongly present in his other novels instead of women as the principal source of threat.

Mildred's absence from the novel *as herself* is important in other respects, too. She appears twice in the novel, but both times in disguise: Once pretending to be Mrs Fallbrook, Lavery's landlady, and again, at the end, when she faces Marlowe as Crystal and Marlowe confronts her with her real identity. In a key scene when Marlowe meets Mildred in Lavery's house, both play a game with names and identities; Marlowe introduces himself as a bill collector while Mildred pretends to be the landlady. She also maintains an admirable level of cool collectedness given that, only moments before, she killed Lavery and left his body in the bathroom. Mildred's determination and fearlessness are contrasted against Marlowe's discredited position. This scene suggests that the language used by Chandler to describe his detective in the novel is itself at odds with his illustration in *The Simple Art of Murder*, where he claims that Marlowe is

---

[9] Malmgren (93) examines the genre's adherence to romance through the motif of quest. This discourse is related to the treatment of Marlowe as "knight" and women as "heroines." From its very title, Malmgren maintains that *The Lady in the Lake* suggests that the detective's job is to search for and rescue a missing "heroine," Crystal Kingsley, but the search is rendered a failure. Murder becomes meaningless, "less an act than a reaction, unmotivated and incalculable," while the knight, Marlowe, "serves mainly to discover the bodies produced by his own quest for the truth" (95).

"the hero, he is everything" (14). Rather, the dialogue positions the woman as the dominant one and in control:

> I kept on looking at the gun and not screaming.
> The woman came close. When she was close enough to be confidential she pointed the gun at my stomach and said:
> "All I wanted was my rent. The place seems well taken care of. Nothing broken. He has always been a good tidy careful tenant. I just didn't want him to get too far behind in the rent."
> A fellow with a kind of strained and unhappy voice said politely: "How far behind is he?" (LL 77-8)

While Mildred pulls the gun on Marlowe, who is disempowered and helpless, she also invades his space when she comes 'close enough' to him, addressing him in an assertive tone, pretending to have the authority as the landlady. Marlowe's meek response – as he talks about himself in the third person and seems separated from himself – exemplifies the detective's lack of agency, which further underscores Mildred's power.

Mildred's second disguised appearance, as Crystal, occurs near the end of the novel. As in the first instance, Marlowe has also assumed a false identity – this time, of Derace Kingsley. On the first reading of the novel, we assume that the woman is Crystal and her apparent villainy is reinforced. On a second reading, however, the exchange between the criminal femme fatale and the detective appears highly ironic because the position and expectations of the reader shift, having learnt that the woman is not only not Crystal, but is indeed her murderer. When the detective recognises that the woman before him is a con artist and confronts her, he admits that he did not do his job as a detective and that he himself was wrong about his judgments:

> "You do this character very well," I said. "This confused innocence with an undertone of hardness and bitterness. People have made a bad mistake about you. They have been thinking of you as a reckless little idiot with no brains and no control. They have been very wrong" (*LL:* 148-9)

Marlowe's acknowledgment that Mildred is not "a reckless little idiot" and that people "have been very wrong" about her is testimony to her agency. In this final confrontation, Mildred jeopardises the detective's power with her gun for the second time: He is threatened physically and his masculinity suffers under this threat:

> …he threw the words out, almost with a lilt. "You wondered why I gave you the empty gun. Why not? I had another one in my bag. Like this."… I

grinned. It may not have been the heartiest grin in the world, but it was a grin. (*LL* 149-50)

Marlowe's powerlessness is further reinforced at the end when Degarmo interferes and hits Marlowe with a blackjack; the latter wakes up to find Mildred murdered.[10] Here, Chandler establishes a connection between the criminalised and sexualised body of the woman on the one hand and Marlowe's vulnerability and lack of agency on the other. The scene depicts Marlowe as perplexed at what has happened to him; he does not know where he was and how Mildred's body ended up next to him. Marlowe vividly describes losing control over his body: "The gin was in my hair and eyebrows, on my chin and under my chin… My coat was off and I was lying flat on my back beside the davenport on somebody's carpet and I was looking at a framed picture" (*LL* 152). This image of Marlowe who "groaned, and made a grunt out of the groan, from professional pride or what was left of it" is important as it mirrors Mildred's dead body: "I looked around for her. She was still there. She was lying on the pulled-down twin bed (*LL* 152-3).[11]

Mildred's naked dead body is paralleled by the vivid description of Marlowe's immobility and his lack of control. Chandler does not give his detective the upper hand even here. Mildred is dead and lying on the floor, he is alive but powerless.[12] The 'big sleep'[13] – death – at the finale of this

---

[10] Marlowe is drugged and beaten up on more than one occasion in the novel. The first time is when the police force him to drink alcohol, arrest him and beat him up, and more importantly in the final scene of the novel. He often loses consciousness and his state of unconsciousness, as Woody Haut suggest, is symmetrical with the era's political apathy and lack of power (74). He reverts to humour and wisecrack when he faces situations that he has no control over.

[11] A number of scholars contend that Chandler's fictions pay close attention to the male body. But readings such as that of Gill Plain, for example, question Marlowe's control suggesting that Marlowe tries to assert his masculinity through the "protection of the weak, both male and female, and through a sentimental, paternalistic romanticism that stands in stark contrast to the isolated existentialism of the tough guy persona" (60-1). Women in Chandler's writing, according to Plain, are devoid of sensuality and are "harsh and unforgiving, delineat[ing] female sexuality that is perceived as threatening even as it attracts" (61). See also Forter for an interesting examination of hardboiled masculinity, especially in the context of masochistic pleasure.

[12] Hiney describes Marlowe in this novel as "overshadowing other characters too much, and caring too little" (131). This story therefore lacks tension because Marlowe is so disengaged and cynical that the outcome seems barely relevant to him. Robert Merril also contends that the lack of engagement in this story accounts for its reception as one of Chandler's weaker works (9).

novel, manifests the sexual signification of the woman's body. Mildred "was wearing a pair of tan stockings and nothing else. Her hair was tumbled. There were dark bruises on her throat" (*LL* 153). The sexual nature of Mildred's murder indicates that Chandler treats femininity as a source of threat, and that her death is a consequence of her dangerous femininity: "Across her naked belly four angry scratches leered crimson red against the whiteness of flesh. Deep angry scratches, gouged out by four bitter fingernails" (*LL* 153). Chandler's description of Mildred's body sexualises her even in death, the descriptions of her 'naked belly' and 'angry scratches' alluding to the sexual nature of the crime perpetrated against her. However, the way her death is presented alongside Marlowe's powerlessness problematises a simple reading that focuses merely on the containment of the woman's body.[14] Mildred's death places her out of Marlowe's reach and, thus, endows her with power even in her total absence from the scene. She is, thus, neither controlled nor contained.

The ending of the novel also unsettles the normative stability of law and order through the death of the corrupt cop whose authoritative agency is transferred to Mildred through her criminalisation. Indeed, *The Lady in the Lake* draws a parallel between the corrupt policeman and Mildred's femme fatality. Hence, Chandler's use of parallels in the post-mortem unification of Degarmo and Mildred as well as the exchanging of the female characters' identities discussed above provides a critique of legal and medical institutions. *The Lady in the Lake*, thus, reflects wartime disillusionment with a thoroughly corrupt system. But if one can say anything about these faked identities and complexities in demarcating the lines between the criminal and victim in this novel, it is the multiplicity of what Peter Rabinowitz calls, the 'evil doers' (129) in this novel and the unity of their gender as women. And although the two women of the book end up dead, Chandler builds up a narrative that invites a reading of women's absence which points to their agency and power, ultimately deconstructing the legendary hardboiled persona of his detective 'hero'.

Indeed, Chandler's narrative carries a transgressive potential personified in the figure of the criminal femme fatale who opens up a multitude of possibilities to reread gender roles in this genre of crime

---

[13] This is a reference to Chandler's first novel entitled The *Big Sleep* in 1939.

[14] For example, William Marling (102) makes the argument that sexual promiscuity is the source of all troubles in Chandler's work, a danger that must be contained and suppressed. Stephen Knight also argues that Chandler's treacherous women reveal "a deep gendered set of evaluations at the core of his novels" and that "sexist violence" of his fiction is euphemised by the wit and vitality of the texts (*Crime Fiction* 120).

fiction. My argument, which offers a new perspective on Chandler's female characters that departs from the critical scholarship on this author, redefines this position of the femme fatale within the genre of hardboiled crime fiction. As I have shown, the criminal femme fatale goes beyond a clichéd character in a genre that celebrates tough masculinity to occupy a powerful space at the heart of the narrative. This space grants the criminal femme fatale the power, freedom and fluidity to disrupt and to break her own stereotype.

Works Cited

Abbott, Megan. "'Nothing You Can't Fix': Screening Marlowe's Masculinity." *Studies in the Novel* 35.3 (2003): 305-324. Print.

—. *The Street was Mine: White Masculinity in Hardboiled Fiction and Film Noir*. New York: Palgrave Macmillan, 2002. Print.

Chandler, Raymond. *The Lady in the Lake*. London: Pan Books, 1959. Print.

—. "The Simple Art of Murder." *The Second Chandler Omnibus*. London: Book Club Associates, 1973. Print.

Durham, Philip. *Down These Mean Streets a Man Must Go: Raymond Chandler's Knight*.

Durham: U of North Carolina P, 1963. Print.

Forter, Greg. *Murdering Masculinities: Fantasies of Gender and Violence in the American Crime Novel*. New York: New York U P, 2000. Print.

Grossman, Julie. *Rethinking the Femme Fatale in Film Noir: Ready for Her Close-up*. Basingstoke: Palgrave Macmillan, 2009. Print.

Hanson, Helen. *Hollywood Heroines: Women in Film Noir and the Female Gothic Film*. London: I.B. Tauris, 2007. Print.

Haut, Woody. *Pulp Culture: Hardboiled Fiction and the Cold War*. London: Serpent's Tail, 1995. Print.

Hiney, Tom. *Raymond Chandler: A Biography*. New York: Atlantic Monthly P, 1997. Print.

Horsley, Lee. *Twentieth-Century Crime Fiction*. Oxford: OUP, 2005. Print.

Kaplan, E. Ann, ed. *Women in Film Noir*. London: British Film Institute, 1998. Print.

Knight, Stephen. *Crime Fiction 1800-2000: Detection, Death, Diversity*. Basingstoke: Palgrave Macmillan, 2004. Print.

—. "'A Hard Cheerfulness': An Introduction to Raymond Chandler." *American Crime Fiction: Studies in the Genre*. Ed. Brian Docherty. Basingstoke: Macmillan, 1988. 71-87. Print.

Lid, R. W. "Philip Marlowe Speaking." *Hardboiled Mystery Writers: Raymond Chandler, Dashiell Hammett, Ross Macdonald: A Literary Reference.* Eds. Matthew J. Bruccoli and Richard Layman. New York: Carroll and Graf Publishers, 1989. 37-56. Print.

Malmgren, Carl D. *Anatomy of Murder: Mystery, Detective, and Crime Fiction*. Bowling Green: Bowling State U Popular P, 2001. Print.

Marling, William. *Raymond Chandler*. Boston: Twayne Publishers, 1986. Print.

Maxfield, James. The *Fatal Woman: Sources of Male Anxiety in American Film Noir, 1941-1991*. Madison: Fairleigh Dickinson U P, 1996. Print.

Merrill, Robert. "Raymond Chandler's Plots and the Concept of Plot." *Narrative* 7.1 (January 1999): 3-21. Print.

Mulvey, Laura. "Visual Pleasure and Narrative Cinema." *Media and Cultural Studies: Keyworks.* Eds. Meenakshi Gigi Durham and Douglas Kellner. Oxford: Blackwell, 2006. Print.

Nyman, Jopi. *Men Alone: Masculinity, Individualism, and Hard-Boiled Fiction.* Amsterdam: Rodopi, 1997. Print.

Plain, Gill. *Twentieth-Century Crime Fiction: Gender, Sexuality and the Body*. Edinburgh: Edinburgh U P, 2001. Print.

Rabinowitz, Peter J. "Rats Behind the Wainscoting: Politics, Convention, and Chandler's *The Big Sleep*." *Critical Response to Raymond Chandler*. Ed. J. K. Van Dover. Westport: Greenwood P, 1995. 117-137. Print.

Skenazy, Paul. *The New Wild West: The Urban Mysteries of Dashiell Hammett and Raymond Chandler*. Boise: Boise State Western Writers Series, 1982. Print.

Speir, Jerry. *Raymond Chandler*. New York: Ungar, 1981. Print.

CHAPTER FOUR

THE DESI WIFE IN AMERICA: REPRESENTATION OF DOMESTIC SPACES IN THE FICTION OF BHARATI MUKHERJEE AND CHITRA BANERJEE DIVAKARUNI

SUMANA GUPTA

Women's writing in the West is shaped primarily by the concept of gender discrimination and is largely preoccupied with women's rights, the reclamation of women's texts written in the nineteenth century and reinterpreting them. In Third World countries, particularly in the Southeast Asian sub-continent, women's writing encompasses a long history of struggle for women's emancipation and education. While the history of the women's movement in the world saw different stages of development chronologically since the nineteenth and early twentieth centuries through the first, second and third wave of feminism, in India, women's entry into nationalist politics during the early twentieth century marked the Indian woman's liberation from the boundaries of home and entry into the larger arena of involvement in social and national issues. While the Western concept of feminism stressed individual rights and was concerned with the improvement of women's status in society, the Indian chapter of women's history still clung to the feminine virtues of a woman – the repository of the traditional culture of modesty, meekness and the need to be protected by men. Women in India and the rest of the sub-continent enjoyed less freedom of expression than their Western counterparts. They were socially constructed as voiceless and when they were legitimised to speak, they were to speak in the voice given to them by men.

Since the 1960s, the Indian diaspora became an area of academic interest as more and more writers from the Third World made their presence felt in the larger literary scenario which had hitherto been dominated by writers from the West. Post 1965, US legislation allowed migration of families from Asian countries and female immigrants started making their presence felt through the memoirs and narratives in which a woman subject could express herself publicly. Sandhya Shukla, in her seminal book *India Abroad: Diasporic Cultures of Postwar America and England*, observes that the first Indian authors to have claimed US literary celebrity were female. These woman writers who dominated the literary scene between the 1970s and 1980s in the United States include Bharati Mukherjee, Chitra Banerjee Divakaruni and Meena Alexander.

Writings by Third World women of the South Asian diaspora focus on the double marginalisation of women – within the patriarchal social structure and as a dislocated individual trying to relocate subjectively in an alien culture. The diaspora experience happened to Indian women post-independence during the 1950s and 1960s. These dislocations were partly due to personal reasons as many Indian wives followed their husbands to the US leaving their families in the home country. These women were often beset with an overpowering guilt at leaving the home culture behind and therefore, initially resisted any form of assimilation into the host culture, trying to maintain and preserve the sanctity of the traditions that she carried from the home country. In being uprooted from the original homeland and replanted in a different soil and culture, these women fought the trauma of dislocation and were faced with the anxiety of preserving their native culture in the alien land. The traditional Indian wife became a stereotype of the role for which she had been conditioned in her *desh* and which now weighed heavily as she began to be considered the repository of tradition in *pardesh* – the alien country of dislocation.

Women are the subjects in whom the effects of migration are most visible due to their location at the centre of the cultural hub, the home – the site of shifts and changes. Memories of the 'original' home continue to haunt the exile-like spaces that a woman dislocates to and wants to make a home. Memories of the traditions, culture and language that she tries to superimpose on the 'real' living in the host country, creating a replica of the home she has left behind, does not necessarily corroborate with the real homeland but to the desire of replicating, reconstructing a home in the diaspora. Such women are represented as individuals who exercise their choice in either conforming to traditional roles or in questioning the convention and asserting their individuality. The emergence of the new consciousness changed the language of representation as women began to write their own histories in their narratives, articulating the differences in subject position as woman writers, breaking the silence and giving a voice to their own history of struggles and pain. In the writings of Bharati Mukherjee and Chitra Banerjee Divakaruni, the twin burdens of being Indian immigrant women and struggling to survive the change are evident.

Though Mukherjee had started writing and publishing much before Divakaruni, the same strains of dislocation and marginalisation are visible in the writings of both writers. Mukherjee had her own theory of immigration; she has discarded her Indian identity to embrace an American identity. Yet, in some of her short stories and novels, she faithfully portrays the dilemma of young Indian wives coming to America and finding it difficult to assimilate.

The representation of Indian women and their diasporic experiences in Mukherjee's work is quite significant given her own perspective as a woman writer refusing to belong to any 'ism'. In 1992, she said in an interview to Runar Vignisson: "I don't call myself any 'ist' and I don't follow any 'ism'" (162). Mukherjee's women evolve as individuals in the course of their personal journeys and often achieve multi-cultural diversities. Though some women are trapped in cross-cultural exchanges and become victims of violence, they refuse to be dominated or subjected to fixed identities. Mukherjee's second novel, *Wife* (1975), recounts the experiences of Dimple, the Indian wife who accompanies her husband to America and is faced with the dilemma of identity in a land that upholds the virtues of individual liberty. Dimple comes from a middle class Bengali family that considers marriage an important part of a woman's identity. Dimple is intrigued by the superficial paraphernalia of 'becoming' a wife and hence, she tries all available means to make herself beautiful and desirable as a bride. Even before her marriage, she seems to live in a world of fantasy with many expectations and few apprehensions. She gets a reality check when she is married to Amit Basu and is thrust into a domestic space that does not fulfil her romantic dreams; along with a tired husband, her marriage offers her such add-ons as a mother-in-law and a lack of basic amenities in a single room apartment filled with cockroaches. The sense of being dislocated and the feeling of not belonging begin in the life of an Indian wife with marriage and shifting of home to the in-laws'.

Dimple's favorite pastime becomes dreaming of America because the prospect of moving there provides her with a sense of relief, a release from her present surroundings so that she can create a new identity for herself. Her pregnancy comes as a hurdle and she chooses to stage an abortion to be free to move according to her will. Dimple's unhappy marriage proves to be no better on American soil as she continues to grope for a purpose to live and her husband remains unaware of her deranged mind. She contemplates suicide and evolves her own theories of suicide and murder as she thinks there are "seven ways to commit suicide in Queens" (102) and "ten ways to murder" (117). Her psychic condition worsens as she fails to acculturate or assimilate and it degenerates into self-tortuous impulses that her husband fails to understand. As her self-destructive impulses pile up, she resorts to violence and ultimately ends up killing her husband. Dimple lacks the sensibility to encompass the totality of self and remains entangled in her desires; she wishes to see the world around her as she wants it to be, not the way it is.

From exploring the domestic space in *Wife*, Mukherjee moves on to a wider diasporic space in her first collection of short stories, *Darkness* (1985). This collection of stories trace the movement from arrival to settlement as the characters still struggle to come to terms with the trauma of displacement, nostalgia and in-betweenness. In the introduction to the collection, Mukherjee defines her stories as "a movement away from aloofness of expatriation to the exuberance of immigration" (3); yet, in some of the stories, Indian wives are depicted as suffering from similar anguish to be and to belong. In the story "Hindu", the protagonist, Leela Lahiri, aspires to be an American citizen and yet, holds on to her cultural background of being a Bengali woman. Her attempts at assimilation are frustrated because she senses the feeling of being an 'other' in America. She says: "No matter what language I speak, it will come out slightly foreign, no matter how perfectly I mouth it" (140). In another story, "Angela", the protagonist, Angela, is an expatriate from Dhaka who wants to assimilate the American culture to shed the bitter memories of her past; but she agrees to marry an expatriate Indian who holds onto the culture and traditions of the homeland. In the story, "The Lady from Lucknow", the protagonist, Nafeesa Haffeez, is an immigrant wife who has settled in America and the newfound liberty motivates her to indulge in an illicit affair with a white man. She is disillusioned to discover, in the end, that the affair was only a passing interest for the sixty-five-year-old American. In the story, "Visitors", the immigrant wife, Vinita, learns a few things about American culture and easygoing way of life. She had expected to exchange her native culture for an alien one but soon realises that the Indian community in America is more concerned with making money than integrating. The American experience merely enables her to allow strangers to come in and go out of her house during her husband's absence.

Mukherjee faced severe criticism from feminist critics after the publication of *Wife* for presenting a passive female protagonist. Reflecting on this, she said in an interview to Connell and others: "What is regarded as passivity, or was regarded in *Wife* as passivity, by feminist *Ms.* magazine-type readers in 1975, was meant to be read very differently. My women are utilising the tools at hand." (49) Mukherjee wants to depict that the process of acculturation in an alien culture is not an easy one for these women who are excited at the prospect of change but trapped in their own cultural beliefs and traditions. They evolve as they assimilate, learning and growing in the process and finally asserting themselves.

Chitra Banerjee Divakaruni, another well-known writer of the Indian-American diaspora, migrated to the US in 1978 as a scholar. She gradually became involved with Afghan women refugees and other abused women in the US. Her counseling sessions with these abused women opened up a Pandora 's Box as she gathered information on the problems of immigrant women in the US and started addressing these issues in her writing. Her primary focus became the cross-cultural experiences of women in America – women battling the prejudices and discriminations in an alien society – and the conflict between the old and the new world orders. Her women subjects find America a mixed blessing because the American experience can be both liberating and discriminating. If her immigrant women protagonists celebrate immigration as a liberating agent freeing them from the old customs of home, they also tackle the challenges of dislocation, isolation and racial discrimination. Divakaruni started writing about women from the South Asian subcontinent – Indians, Pakistanis and Bangladeshis – in her first volume of poems, *Black Candle* (1991). The women depicted in this book face oppression from within the family by being forced into marriages that don't work or suffering in the husband's extended family for childlessness. They are abused, beaten and often compelled to share their husband with a second wife.

In her next book, *Arranged Marriage* (1995), which is a collection of short stories, she began focusing on women battling abuse and suffering and 'coming out triumphant'. The 11 stories in the collection deal with marital relationships and, more importantly, speak about a culture in which the institution of marriage is 'arranged' by consenting parties. Coming from Bengal and imbibing the Bengali culture, Divakaruni has portrayed her own imaginings and the experiences of others in these stories that deal with domestic relationships, violence, racism, cross-cultural interactions, liaisons, shame, abortion and separation. The stories in the book are based on two major premises. A marriage that is arranged according to the customs of society does not guarantee happiness or unquestioning loyalty to the revered institution and the immigrant experience is not always filled with hope because there are challenges of discrimination and marginalisation. If the old customs and beliefs stifle, the newfound liberation does not always promise happiness. The woman is often faced with the choice of conforming to tradition or embracing modernity and, in exercising her choice, she voices her individuality. The Third World woman that is constructed in these stories of immigration is different from Western feminist representation that sees Third World women as monolithic constructs – mostly domesticated housewives who are uneducated, voiceless victims of abuse as opposed to the educated, liberated women of the West. Divakaruni wants to show that in conforming to traditions or choosing to be modern, the Third World woman exercises her self-will; she has control over her own body and sexuality and takes her own decisions.

The story which most poignantly reveals that the American dream is not without its shocks is "Silver Pavements, Golden Roofs". In this story, Jayanti, a young Indian college girl who comes to America with the illusion that it is a land where promises are fulfilled, is disillusioned when she encounters racial abuse on the streets of Chicago and witnesses domestic violence against her aunt by her unsophisticated uncle. She declares: "No arranged marriage like Aunt's for me" (45). But Jayanti goes on to realise that the American experience is not simple and she will have to accept both its beauty and its pain in order to survive. In choosing to let go of her illusions and preparing to battle it out, she shows the promise of arrival and survival. In another story, "Doors", the woman protagonist, Preeti, brought up in a Westernised environment, decides to marry an Indian man against her mother's wishes. In this love marriage, the husband, with his own set of traditional values, fails to appreciate Preeti's love of the idea of privacy and closed doors and Preeti chooses to express her discontent by questioning the traditional values imposed upon her within the institution of marriage. "The Disappearance" is the story of an immigrant wife from Calcutta who chooses to walk out of the marriage for no obvious reason that the husband can recognise. He has only exercised his normal authority as the master of the house on his wife who is also a part of his property. For him, his wife has been a symbol of his financial stability and sexual power. He is so completely constructed by patriarchal values that he fails to understand why his wife has disappeared with her jewellery. In leaving the husband, the wife has transgressed the normative behaviour of a traditional Indian wife who is not expected to leave her husband upon any provocation. In the story, "Clothes", the wife, Sunita, chooses to stay back in America after her husband dies rather than going back home to serve her in-laws: "I know I cannot go back. I don't know yet how I will manage – here in this new, dangerous land. I only know I must. Because all over India at this moment, widows in white saris are bowing their veiled heads, serving tea to in-laws. Doves with cut-off wings!" (33) In choosing Western clothes and America, she chooses steadiness of resolution over apprehension. Divakaruni has drawn parallels between immigrant women and women at home in two stories, "The Bats" and "The Maid Servant's Story", where the women subjects are dominated and manipulated to remain in abusive marriages to put up appearances in the society. They choose to stay in the spaces of home which have the potential of helping combat patriarchal oppression by fostering new bonds between women – bonds that resist patriarchal domination. In contrast, the immigrant woman is doubly marginalised – within the family and as a minority subject in an alien culture. In the story, "Affair", Abha refuses to put on Western clothes or wear make-up to please her husband but it is only through Meena that she realises the fruitlessness of her marriage and the need to leave a futile relationship. In extricating herself from a mismatched marriage, she charts her own route of survival in America. In "Ultrasound", which Divakaruni later developed into a full-length novel, *Sister of My Heart*, the bonding between two women, Anju and Runu, over the fight against abortion of a female foetus shows the cracks in their respective marriages. If Runu's husband and his family are more orthodox and Anju's husband, Sunil, apparently more liberal, the undertones of Sunil's arguments against Anju persuading Runu to come to America are more disturbing as his words seem to imply to Anju: "See how lucky you are to have a husband like me, to live in this free and easy American culture… You'd better start working harder at being a good wife." (218) In the last story, "Meeting Mrinal", the image of perfection that Asha had built in reconstructing her life in America after divorce proves to be a fake and leaves her with a sense of inadequacy as she goes to meet her friend Mrinal who, she thinks, 'has the perfect existence'. The very images of perfection against which she has chosen to measure herself collapse as Mrinal breaks down in tears to admit that she is unhappy and lonely and envies her. Asha's images of perfection that had sustained her now become her torment. She wonders: "What would I live on, now that I knew perfection was only a mirage?"(296) It is only through letting her son, Dinesh, take care for her and letting go of the perfect yet unhelpful images of the ideal wife and mother that she can raise a toast to "precious, imperfect lives"(300) that are to be lived without illusions, appreciating their infinite possibilities.

Most of the stories in this collection are told from the first person point of view that connects with readers immediately, creating a bond of intimacy, a sharing that is both personal and universal. The institution of marriage, which has long been cherished in India as sacred and fulfilling, has eroded with time to reveal cracks and holes. Divakaruni was criticised for tarnishing the image of India and the Indian community by presenting negative images but in this book, her aim was to shatter stereotypes. She wrote a regular column, 'Spice of Life' for the online magazine, *Salon.* In a column titled 'My Fictional Children' she wrote:

> My writing is made more complicated by the fact that I'm exploring the experience of being Indian, of being brought up in a culture where many still consider motherhood a woman's supreme destiny, and the inability to get pregnant her supreme failure (28 Jan, 1998)

She herself insists on maintaining her traditional culture ("The Indian part of my culture is very important to me") but also admits that sometimes, the feeling among the Indian community that she is betraying the community by pointing out its problems is pretty disturbing. She voices her concern in an interview to Neela Banerjee of *Asian Week.com*:

> You are expected to be a spokesperson for the community, and that is just an unfair kind of burden. I always try to make it clear that I am presenting one vision about what is true about the Indian-American community. It is a very diverse community, and mine is just one angle of looking at it (27 April-3 May 2001).

Divakaruni wanted to convey through her women characters that within the institution of marriage, women are circumscribed by traditions that define their identities. When they choose to break free of the roles, they offer a resistance to the patriarchal structure of domination within the family and the diasporic space allows the woman subject to construct an identity and a room for herself where she can voice her differences and articulate her desires.

Works Cited

Connell, Michael, Jessie Grearson, and Tom Grimes. "An Interview With Bharati Mukherjee." *Conversations With Bharati Mukherjee*. Ed. Bradley C. Edwards. Jackson: U P of Mississippi. 2009.18, 37-58. Print.

Divakaruni, Chitra Banerjee. *Arranged Marriage*. Black Swan: UK. 1997. Print.

—. Interview to Neela Banerjee of *Asian Week.com*. 27 April-3 May 2001. Print.

—. "My Fictional Children". *Salon.* 28 Jan, 1998. Print.

Mukherjee, Bharati. *Wife.* 1975. New Delhi. Penguin. 1990.

Shukla, Sandhya. *India Abroad: Diasporic Cultures of Postwar America and England*. New Delhi: Orient Longman. 2005. Print.

CHAPTER FIVE

"INDIAS OF THE MIND": REPRESENTATIONS AND MISREPRESENTATIONS IN SISTER NIVEDITA'S WEB OF INDIAN LIFE AND KATHERINE MAYO'S MOTHER INDIA

DEBALINA BANERJEE

> Hinduism does not speak of women with one voice… Hindu thought has thrived on controversy and contradiction, perhaps more so on the subject of women than on any other… It is through ambiguities and paradoxes that the Hindu tradition projects its conception of women (Bose 10).

I remember reading somewhere that a lady had once remarked to Webster, "Mr Webster, you have put some naughty words in your dictionary," to which Webster had simply bowed and replied, "Madam, you looked for them." Mayo's and Nivedita's representations of India remind me of this particular anecdote. Interestingly, Mayo's *Mother India* reminded Mahatma Gandhi of a drain inspector's report. While both texts remain constructions for the Western audience, they represent a different India that they purportedly 'looked for' and subsequently came to understand, internalise and project. And reading them together gives rise to an interesting dialectic and dialogue called India.

The Indian struggle for Independence was bolstered by invocations to the cult of *shakti*. Bankim Chandra Chattopadhay's *Anandamath* popularised 'the nationalist iconography of the nation as Mother India' (Sinha 625). Partha Chatterjee, in his famous critique, "Nationalist Resolution of the Women's Question in Nineteenth Century Bengal", focuses on the gendered discourse of Indian nationalism. It is as if the nationalists, to corroborate their demand for self rule, chose the woman question as the contested ground for settling these unsettling questions. As opposed to negative projections of Indian womanhood, the need to 'refashion' the identity of the Indian woman was seen as the need of the hour. Hence, the emergence of the educated *bhadromahila*, poised and genteel, in sharp contrast to her unlettered and unrefined counterpart, yet grounded to the demands of home and the hearth. The woman question was the one sphere the nationalists refused to negotiate with the colonial state and was hence beyond the domain of imperialist control. Her difference from the Western woman was the obvious and most agreeable outcome of this project of modernity. However, "This neglect of women's agency in the outer world of nationalist politics not only underestimates the discourse of Indian feminism in the early women's movement, but also freezes the gendered logic of Indian nationalism to a single movement supposedly defined by the singular problematic of the assertion of cultural 'difference' from the West"(Sinha 625).

When American journalist Katherine Mayo, in 1927, wrote her infamous 'polemic against Indian self-rule', interestingly and quite understandably, she called her book *Mother India*. Mayo herself observes in her other vitriolic book on India, *Slaves of the Gods,* that she had chosen the title deliberately to awaken the women of India by contrasting the actual treatment of women with their glorification in nationalist discourses[1]. This kind of cultural imperialism tends to rob women of their local, historical, religious and cultural contexts. According to writer Chandra Talpade Mohanty, Western feminists base their understandings of their Third World counterparts on internalised notions of 'racism' and 'classism'. Third World feminists object to portrayals of women of non-Western societies as passive and voiceless victims as compared to the portrayal of Western women as modern, educated and empowered. Mayo's goal was to uphold British imperialism in India, pulverising the nationalist politics.

> From every political platform stream flaming protests of devotion to... Mother India; but India's children fit no action to their words. Poor indeed she is, and sick, ignorant and helpless. But, instead of flinging their strength to her rescue, her ablest sons, as they themselves lament, spend their time in quarrels together or else lie idly weeping over their own futility... Meantime the British Government, in administering the affairs of India, would seem to have reached a set rate of progress, which, if it be not seriously interrupted, might fairly be forecast decade by decade. (Mayo 12)

Significantly, Katherine Mayo's *Mother India* begins with an account of Hindu religious life as seen through the myopic lenses of the coloniser,

---

[1] Mrinalini Sinha. "Refashioning Mother India: Feminism and Nationalism in Late-Colonial India". *Feminist Studies*. 26:3 (Autumn, 2000). Points of Departure: India and the South Asian Diaspora, 623-644. Print.

burdened with the need to liberate Indians from the morass of ignorance, religious dogma and pagan practices. Kali is described as a Hindu goddess, wife of the great god, Siva, "Black of face she is, with a monstrous lolling tongue, dripping blood" (5). Kali's power is attributed to her lust for destruction, blood and death-sacrifice:

> "Kali! Kali! Kali!" shout all the priests and the suppliants together, some flinging themselves face downward on the temple floor. Meantime, and instantly, a woman who waited behind the killers of the goat has rushed forward and fallen on all fours to lap up the blood with her tongue… "in the hope of having a child." (Mayo 5)

Her Orientalist representations of idolatry play upon commonplace ideas and stereotypes of the Hindu religion.

Even though Mayo's book produced numerous defensive 'patriarchal' and 'indigenous responses' in India, her more serious nationalist critics realised immediately that they could not afford to deny the prevalence of some of the backward social practices that Mayo had found in India (Sinha 628). The backlash at colonial power and its imperialist agencies, typified by people like Mayo, was in the form of organised women's movements. This led to the demand for the passage of the Child Marriage Restraint Act, or the Sarda Act, as it was popularly known[2]. The emergence of movements like the Self-respect Movement called for a radical restructuring of society through the reconstitution of previous gender, caste and class hierarchies. Swami Vivekananda was one of the champions of India's cultural heritage. The nineteenth century was an era characterised by increasing distance between the 'advanced' West and the 'backward' East. The colonisers ruled through systematic debasement of the morale of the native Indian by effectively denigrating India as a civilization and through explicit suggestions about the effeminacy and ineffectuality of the Indian *baboos* and how, in turn, it necessitated control by a higher civilization – in this case the British. "The self-esteem of the Indian educated middle classes was at its lowest ebb by the end of the century. In these circumstances Vivekananda comes and affirms the greatness of India's ancient civilization. It was a turning point in diminishing the native Indians' sense of inferiority" (Bhattacharya 367).

Margaret Elizebath Nobel was an Irish woman who became one of the most illustrious disciples of Swami Vivekanada and came to be known as Sister Nivedita. She dedicated her life to the service of the Indian people, especially women, and thereby became one of "The greatest interpreters of

---

[2] ibid.

ideals, cultures, religions, customs and manners of India" (Introduction *The Web of Indian Life*). Nivedita's purpose, unlike Mayo's, was to instill self-respect in and reorient the demoralised Indian masses, slaving under a hated colonial regime, to the glories of their own ancient civilization. Nivedita often glosses over the unsavory realities of Indian (read Hindu) life but one has to understand her ideology as a step towards the recovery of national glory. *The Web of Indian Life* by Margaret Elizabeth Nobel or Sister Nivedita is clearly one of the earliest manifestos of Third World feminism. Sir Jadunath Sarkar, the eminent historian, while delivering a talk at the Ramakrishna Mission Institute of Culture, Kolkata observes:

> …a nation's chief strength comes out of a belief in its own capacity, and such a belief is the natural product of a conviction of the greatness of its ancestors, the reassuring thought that what they had done their progeny can do. Hence she glorified, and urged us to glory, in ancient India's achievements, not merely in spiritual culture (which is known to all), but also in arts and science, commerce and crafts. She thus became the trumpet voice of what is wrongly called 'Aggressive Hinduism'. (*Bulletin* 2)

This was the time when most 'English-educated Bengalis' looked at the Hindu religion with skepticism and contempt. Either they converted to Christianity or joined the *Brahmo Samaj*. As a backlash, the early 1880s saw the rise of reactionary orthodox Hinduism. And yet, as Sarkar finds, "Nivedita was the furthest removed from the blatant boastfulness of a particular school of noisy, modern, orthodox Hindus. Indeed, she could not be otherwise as a disciple of Vivekananda who was never weary of saying that salvation cannot come from non-touchism (*chut – achut – margi*) but from the heart's strength and purity" (Sarkar 3).

Sister Nivedita, in her *Web of Indian Life*, provides a secular and liberal perspective from which she herself observes, "How happy were those days in the little lane! How unlike the terrible pictures of the Hindu routine which… had embittered my English childhood! Constant ablutions, endless prostrations, unmeaning caste restrictions, what a torture the dreary tale had been! And the reality so different!" (Nivedita 17) Nivedita's dictum that, "Education is vitally determined by circumstances of place", (Nivedita 4) at once places *The Web of Indian Life* within the subversive tradition that defies attempts to chronicle female history in one generalised sweep. Nivedita, tracing the stoicism in Indian women like Bisapala and the Rani of Jhansi, says, "Nor, amongst these strong outstanding types, is there any failure of individual achievement" (Nivedita 7). Mayo, on the other hand, argues that in a majority of orthodox Hindu households, the girl child is, "completely unlettered, her

stock of knowledge comprising only the ritual of worship of the household idols, the rites of placation of the wrath of deities and evil spirits, and the detailed ceremony of the service of her husband, who is ritualistically her personal god" (Mayo 13). And her avowed purpose in life is child-bearing and "matters of procreation are the woman's one interest in life, her one subject of conversation, be her caste high or low". (Mayo 13)

Nivedita, too, was well aware that the lack of education for women was like a canker eating away at the heart of Indian national character. Hence, she began her services to Indian womanhood by establishing the first public school for women in Calcutta. The women, she felt, must aspire to be fit partners to a new race of Indian men in terms of learning and material achievement. While she "praised the practical capacity and power of organisation shown by our elderly housewives at cooking for and feeding hundreds at domestic ceremonies …", it was her distinct desire that "The Indian wife must help her husband in writing his researches". That, for her, was the ultimate 'test' (Sarkar 7). Mayo, talking about the Hindu system of child marriage, observes, "Little in the popular Hindu code suggests self-restraint in any direction, least of all in sex relations" (Mayo 16). She cites the protest of the Brahmin parliamentarian, Shanmukhan Chetty, of Salem and Coimbatore, to give her text the semblance of a dialogue and not make it an European cultural monologue that finds fault with the 'other' – in this case the native Indian's religious customs and traditions: "The fact that a so-called marriage rite precedes the commission of a crime does not and cannot justify that crime. I have no doubt that if you were to ask a cannibal, he would plead his religion for the heinous act he does" (Mayo 21).

In assessing Hindu conjugal life, Nivedita observes:

> In early manhood he trusts to her advice to moderate the folly of his own rasher inclinations, in old age he becomes as everywhere in the world, more entirely the eldest of her bairns, and she more and more the real head and centre of the home. But always she remains as she was at the beginning, Lakshmi, her husband's Goddess of fortune (Nivedita 47)

While Nivedita refers to "the characteristic emotion of the wife" as "passionate reverence" and that of the Hindu husband as "a certainly measureless protection" (Nivedita 45), Mayo cites Miss Cornelia Sorabji's *Between the Twilights* to highlight the woman's subordinate status in the Hindu family hierarchy:

> When she is the mother of a son, greater respect is hers from the other women in the *zenana*… she has been successful, has justified her

> existence. The self-respect it gives the woman herself is most marked. She is still a faithful slave to her husband, but she is an entity, a person, insofar as that is possible in a Hindu *zenana*… (Mayo 38)

Motherhood is a profound form of self expression and a reassertion of female identity. Hence Nivedita surmises, "With the coming of her first-born, be it boy or girl, the young wife has been advanced, as it were, out of the novitiate. She has become a member of authoritative circle. It is as if the whole world recognises that henceforth there will be one soul at least to whom her every act is holy" (Nivedita 21).

Mayo, on the other hand, while continuing her tirade against Hindu religious sanctions, states, "The reverse of the picture shows the Hindu widow – the accursed" (Mayo 40). Mayo borrows from Miss Sorabji her idea of Hindu widowhood replete with hardship and suffering. The widow functions as the 'menial' in her late husband's household, straddled with the hardest and ugliest tasks:

> She may take but one meal a day and that of the meanest. She must perform strict fasts. Her hair must be shaven off. She must take care to absent herself from any scene of ceremony or rejoicing, from a marriage, from a religious celebration, from the sight of an expectant mother or of any person whom the curse of her glance might harm" (Mayo 40).

The choice of words is suggestive of a strong Christian bias. Mayo sees in the Indian widow an unquestioning internalisation of patriarchal dictates, "she herself is the priestess of her own misery, for its due continuance is her one remaining merit" (Mayo 40). Literature, as Nivedita finds it, "consists largely of man's praise of women in relation to himself" (Nivedita 52). Her sentiment echoes the basic tenets of feminism. In this connection, she observes how "Neither Europe nor modern spirit can claim the glory of having created the idea of woman as an individual", "the woman of solitude, the woman who stands alone" (Nivedita 52, 53). Such persons in the world of Indian women are the widows, who receive the sanction of religious life, "Life ebbs but discipline gathers its perfect fruit, in lives stately and grave and dignified for all their simplicity and bareness;… in an ideal of sainthood justified; an opportunity of power created" (Nivedita 62, 63). There is no doubt that Nivedita is sugarcoating a bitter pill as the abysmal condition of Hindu widows was a well-established truth. In trying to address the basic philosophical tenets of Hindu family life, Nivedita, perhaps, is suggestive of what Syam Sundar Chakravarty argues in his treatise, *My Mother's Picture*: *An Attempt to Get at the Hindu Spirit in Connection with the Mayo challenge,* "the evils to which Mrs Mayo refers… insofar as they exist – are really the outcome, not of the orthodox Hindu creed, but of a fall from the creed" (Albinia 432).

Mayo postulates that remarriage in orthodox Hinduism, is impossible because "Marriage is not a personal affair but an eternal sacrament. And it must never be forgotten that the great majority of the Hindus are orthodox to the bone" (Mayo 41). This was also the time when several reformers like Ishwar Chandra Vidyasagar and Raja Rammohan Roy were fighting tooth and nail for the possible emancipation of women, in terms of women's education and widow remarriage. Though she acknowledges the fact that, "In different sections of India, several associations have sprung up, having the remarriage of virgin widows as one of their chief purported objects (Mayo 41)", she is quick to point out that, "The movement, however, is almost wholly restricted to the most advanced element of Hindu society, and its influence is, as yet, too fractional appreciably to affect statistics" (Mayo 41, 42). Her imperialist agenda is unmistakable when she states "Of recent years, however, the gradual if unrecognised influence of Western teaching has aroused a certain response" (Mayo 41).

In his essay on "Modern India", Swami Vivekananda writes: "Have we not… to learn anything from the West? Are we perfect? Is our society entirely spotless, without any flaw? Yes, learn we must many things from the West" (Bhattacharya 369). Yet he was aware that learning from the West should not be mere imitation of the West and alienation from one's own people. Thus, he recommended that Indians, from the highest to the lowest, should retain their identity and sense of belonging and, at the same time, be ready to learn what is worthwhile from the West[3]. Similarly, Sister Nivedita was also aware of the fact that Indian women should be spoken to in the language of Indian ideology and not in terms of the West. According to Sister Nivedita, emancipation would be attainable only,

> When the women see themselves in their true place, as related to the soil on which they live, as related to the past out of which they have sprung; when they become aware of the needs of their own people, on the actual colossal scale of those needs; when the mother-heart has once awakened in them to beat for land and people, instead of family, village, and homestead alone… (Nivedita 95, 96).

While talking about the pangs of labour and the pattern of childbirth in Hindu households, Mayo resorts to distortions and sensationalisation of

[3] Sabyasachi Bhattacharya. *Swami Vivekananda's Impact on B. G. Tilak and the Nationalist* Movement. Web. http://www.sriramakrishna.org/bulletin/b810his.pdf.

details that would aptly suggest to the Western readership the rationality behind her claims:

> When the pains begin, they send for the *dhai*... She thrusts her long-unwashed hand, loaded with dirty rings and bracelets and encrusted with untold living contaminations, into the patient's body, pulling and twisting at what she finds there. If the delivery is long delayed and difficult, a second or a third *dhai* may be called in, if the husband of the patient will sanction the expense, and the child may be dragged forth in detached sections – a leg or an arm torn off at a time. (Mayo 45)

The women at the *purdah* hospital in Lucknow are similarly dehumanised by Mayo as "alike looking out from behind a common veil of helplessness and pain" (Mayo 27). Fran Hosken writes, "Rape, forced prostitution, polygamy, genital mutilation, pornography, the beating of girls and women, *purdah* (segregation of women) are all violations of basic human rights" (Mohanty 346). Mohanty observes that by equating *purdah* with rape, forced prostitution and domestic violence, Hosken "...asserts its sexual control function as the primary explanation for *purdah*. Institutions of *purdah* are thus denied any cultural and historical specificity..." (Mohanty 66). Mayo is in similar denial, expressing a similar sentiment from the vantage point of Hindu cultural practices. Nivedita says, "Even queens in the East are too sacred to be looked upon by common eyes... The long silken tent through which such ladies move from palace-door to carriage-step is no vulgar prison, but a shrine. Bereft of its concealment, they would feel dishonoured..." (Nivedita 50). Here, Nivedita is questioning the importance of choice or will in perpetuating the system of *purdah*. Mayo, on the contrary, is in denial of any such allowance that might be suggestive of agency among the women of India as it would be antithetical to her aggrandisement of the need for British colonial rule in India. Nothing for her can be more ignoble than garrulous, quarrelsome, ignorant women but she would even welcome such a state of being for the Indian women compared to the state in which she purports to find them. Mayo takes up the civilizing mission of voicing the concerns of the Indian women as they couldn't, supposedly, speak for themselves.

Katherine Mayo's *Mother India* appropriates imperialist articulations of the 'woman question' in the form of "white men [and white woman] saving brown women from brown men" (Sinha), providing ideological legitimacy for British colonial rule as the agent of modernity in India. Valerie Amos and Pratibha Parmar feel, "Feminist theories which examine our cultural practices as 'feudal residues' or label us as 'traditional' also portray us as politically immature women who need to be versed and schooled in the ethos of Western feminism. They need to be continually challenged…" (Mohanty 338). It is such a challenge that Sister Nivedita sets for herself as she writes,

> And in certain parts of the province of Malabar woman is actually in the ascendancy… The term *matriarchal* is more accurate, inasmuch as the husband visits the wife in her own home, and the right of inheritance is through the mother. Thus, far from India's being the land of the uniform oppression of woman by a uniform method, it represents the whole cycle of feminist institutions. There is literally no theory of feminine rights and position that does not find illustration somewhere within her boundaries. (Nivedita 79)

Nivedita's ideology is influenced by Swamiji's belief, "That country and that nation which doesn't respect women will never become great now and nor will ever in future ("Thoughts on Women"). Sister Nivedita, too, like Swamiji, attempts to discover the heart of India, going beyond Western categories and, in the process, realising the true national ideal.

> In India, the mother is the centre of the family and our highest ideal. She is to us the representative of God as God is the mother of the universe… Our God is both personal and absolute; the absolute is male, the personal, female. And thus it comes that we now say: "The first manifestation of God is the hand that rocks the cradle". ("Thoughts on Women")

Works Cited

Albinia, Alice. "Motherhood laid bare: How Katherine Mayo and Manoda Devi Challenged Indian Public Morality". Web. *Sarai Reader 2005: Bare Acts*. 428-435. Print.

Bhattacharya, Sabyasachi. "Swami Vivekananda's Impact on B. G. Tilak and the Nationalist Movement". Web. http://www.sriramakrishna.org/bulletin/b810his.pdf. 20 Sept., 2011.

Bose, Mandakranta. *Women in the Hindu Tradition: Rules, Roles and Exceptions*. London and New York: Routledge, 2010. Print.

Butler, Judith. *Gender trouble: feminism and the subversion of identity*. New York: Routledge. 1999. Print.

Mayo, Katherine. *Mother India.* Web. *Project Gutenberg.* E book no.0300811h.html. 15 Sept., 2011.

Mohanty, Chandra Talpade. "Under Western Eyes". *Third World Women and the Politics of Feminism*. Ed. Mohanty, Chandra Talpade, Ann Russo and Lourdes Torres. Bloomington and Indianapolis: Indiana U P. 1991. Print.

Nivedita, Sister. *The Web of Indian Life*. Mayavati: Advaita Ashrama, 1950. Print.

Sangari, Kunkum and Sudesh Vaid, eds. *Recasting Women: Essays in Indian Colonial History.* New Delhi: Kali for Women, 1989. Print.

Sarkar, Jadunath. "Sister Nivedita". *Bulletin.* Kolkata: Institute of Culture, 2009.1-7. Print.

Sinha, Mrinalini. "Refashioning Mother India: Feminism and Nationalism in Late-Colonial India". *Feminist Studies*. 26:3 (Autumn, 2000): 623-644. Print.

"Thoughts on Women – Swami Vivekananda". Web. http://www.writespirit.net/inspirational_talks/spiritual/swami_vivekananda_talks/thoughts-on-women-swami-vivekananda/. 20 Sept., 2011.

CHAPTER SIX

EDUCATION-KNOWLEDGE AND EMANCIPATION: WOMAN'S JOURNEY TOWARDS MODERNITY AND EQUALITY IN THE CONTEXT OF NINETEENTH CENTURY BENGAL[1]

SARMISTHA DE

> No screens are used on the Continent, in Ireland and in America. I specifically asked the Principal of the Medical College this question and his reply was that these girls sit apart from the male students unscreened in all lecture rooms. During the past three years in this college, they have occupied seats in front of the male students, that is, between them and the lecturers. I have asked the lady students here and they distinctly desire that they should not be screened in any way. (Dr Coates' letter to Directorate of Public Instruction, dated 13th February 1886, letter number 898, General Education Proceeding number B3, September 1897, West Bengal State Archives. Hereafter WBSA)

The above-mentioned administrative correspondence occurred when the admission of lady students in Calcutta Medical College had been initiated. It is interesting and worthy to note that lady students of late nineteenth century across the world were quite confident in receiving lectures on anatomy with male students! Indeed, all over the world, the nineteenth century set a new trend in women's lives. The public space was increasingly opening up before women. This was a fascinating time when women were rediscovering themselves while stepping out into the different fields of public life. Their new identities paved the way for self-

[1] A slightly different version of this article has been published in Bengal Past and Present, Journal of the Calcutta Historical Society, vol.128,2009 pp 29-53 (Woman the Lonely Traveller: Her De-confinement and Journey towards Modernity in Bengal).

evaluation and self-introspection, cutting across the age-old ideas and values of patriarchy. The epoch had finally arrived when women started to see and understand themselves from an unexplored perspective.

The nineteenth century ushered in the *Bengal Renaissance* and India took its first tentative steps towards the creation of a modern nation. This forward movement was also inseparably linked with the de-confinement of women – the *purdah* was breached and women dared to step into the public arena. They received public education, participated in public life as professionals and also as social and political activists. In short, this phase changed the life and status of Indian women who were, till then, regarded as the extension of a reproductive system and could be abused, mutilated and burnt at will while being glorified by the scriptures and worshipped in the form of numerous goddesses in the pantheon. The new trend originated from the abolition of sati in the second decade of the nineteenth century and by the turn of the century, women had already started their professional lives in teaching and medical practice.

This paper will attempt to look at this *change* – this *journey* which Indian women had initiated in order to be equal stakeholders in society and be recognised as such. The Bengali intelligentsia, a part of the gentry and the colonial rulers, were catalysts of this change, knowingly as well as unwittingly, selflessly as well as for self-gratification and glorification. They eventually made major contributions towards this changing role of Indian women who tried to reinvent themselves. This subject has been extensively and intensively investigated. This paper will try to examine the core issue of gender equality and the recognition of equality by presenting a few well-documented incidents. And the issue of equality is invariably linked with the development of an individual woman as a complete person. This may be a good yardstick to measure the progress of a society. And finally, it will examine if society is ready to recognise the *enlightened* and *emancipated* woman as a complete person and whether the woman herself is confident to perceive herself as such and claim her rightful place within the family and without.

With the ascension of the East India Company's economic and political fortunes, Bengal went through rapid and radical changes. The changes affected the men's world more directly and forcefully. For example, Brajendranath De (1852 – 1932) was first educated at home and then sent to one of the new schools that taught English. He obtained a scholarship to study in England where he passed the civil service examination, returned home in glory, and began a career which made it necessary to live away from his traditional home.[2] The new *Bengali baboo* had to adapt to the

[2] See Geraldine Forbes.

paradigms of the new dispensation. Some of them, Madhusudan Dutt being the most famous example of the lot, were swept off their feet by the changes. Others created an outer shell around themselves – the interactions with the new regime were limited to that outer shell whereas within, there was the *traditional core*. The new *Bengali baboo* adapted to the new regime, consorted with the new rulers and wore his Western countenance while he bowed before the family deity and sacrificed buffalos and goats at the altar of *Goddess Kali*.

When Western education was introduced to India, waves of new thinking, reasoning and rationality shook the massive edifice of our society, age-old indigenous institutions and culture. Those who received Western education realised that this journey towards modernity was inevitable as well as desirable for the sake of progress of the individual as well as the nation. But at the same time, most of them were very keen to preserve their indigenous identity. They negotiated with themselves how these two identities could share the space – insulated some space from the world without where the *traditions* were preserved, reinvented and nurtured. This journey towards the 'modern', 'rational' and 'scientific' world peacefully coexisted with 'religion' and 'tradition'. As Partha Chatterjee has pointed out – "Indians pursued science, technology, rational economics and Western political forms while regarding the home as the source of his true identity that needed protection and strengthening and not transformation."[3]

The gale of change found gaps in the *deewar* and parted the *purdah* which insulated women from the outside world – the world of men. It will be grossly inadequate to make the dimensionless statement that the enlightened *bhadrolok* lifted the *purdah* and ushered the womenfolk into the sunny world of modernity. The winds of change which forced the *bhadrolok* to wear his 'modern' countenance also forced him to redefine the countenance of his womenfolk. And to be fair to the *bhadrolok*, it should be pointed out that while he was trying his best to appear the 'worthy subject', the new 'English gentleman' was also evolving and preparing for his new role, the role of a 'superior ruler'. The rulers wanted to create *baboos*, who would ape the *sahibs* and serve the Empire better. In the process, the *baboos* stepped over the invisible *laxman-rekha* and the modern Indian *bhadrolok* was born. The *bhadrolok* wanted to create the new Indian women – *bhadromahilas* – who would be their consorts in their new role. The new Indian woman stepped over the *laxman-rekha* which the *bhadrolok* drew for her to be confined within and took faltering

[3] See Partha Chatterjee.

but defying steps towards being complete persons. The birth and journey of the *bhadromahila* has to be posited in this context.

A complex churning process was taking place in the fledgling Empire. The rulers were anxious to prove that they were a 'superior race'. Ideal relationship between men and women was part of the discourse. James Mill in his *History of British India* (1826) argued that women's position could be used as an indicator of a society's advancement. Enamoured of their 'civilizing mission', the British intelligentsia condemned Indian religions, culture and society because of their rules and customs regarding women – they wanted to 'enlighten' and 'uplift' the natives, at least the section they were in direct contact with. And the section of the natives who embraced the changes and wanted to reform their society wanted their wives and daughters to become their worthy companions. Thus Indian women were granted, albeit in a limited and controlled manner, the privilege of education, particularly Western education. Colonial domination set the stage for change, both in England as well as in India. Mary Carpenter visited India four times between 1866 and 1876 with the specific aim of spreading awareness about female education. In her words,

> A great change is taking place in the views of Hindoo gentlemen and a strong desire exists in the mind of the most enlightened among them that the future wives and mothers of their nation should be enabled by sound education to better fulfil the duties of their important spheres.[4]

It was felt that modern education had to be imparted to women so that they could perform the roles of wives and mothers of a modern generation. Thus the 'educational conference' held in 1887 framed a new set of syllabi and standards of examination for girls' schools and (institutions for) *jenanas*. Domestic economy, *garhasthya swasthya bidhi* (domestic health and hygiene) and needlework were incorporated into the syllabus for girls' schools apart from the usual subjects taught to boys – reading, writing, arithmetic, grammar, geography and history. After a separate syllabus for girls was framed, adherence to that was insisted upon. Pupils in girls' schools outside Calcutta "should not be debarred from learning the ordinary course for boys' schools and competing for the ordinary scholarships; but effort should be made to introduce in all such schools the separate standard for girls now recommended".[5]

---

[4] 'Mary Carpenter's notes on female education,' DPI, Bundle -2, File A-1868, WBSA.

[5] Educational Conference 1887, DPI, Bundle – 27, File 210-1893,WBSA.

A section of the Bengali *bhadrolok* started a new discourse on family in the late nineteenth century. The prevailing paradigm of the family was inheritance of property and religious obligations. Within the new paradigm, family becomes an affective and sentimental unit and affinity and 'private intimacy' among the members was the bedrock of such a family. Providing the child with proper education was one of the most important tasks of parents. The attempt to break away from the old moorings was conscious and deliberate. While lessons on the scriptures and the epics were given to children, particularly the boys, English language and Western education occupied the most important place and a degree earned from England or Europe was the most cherished dream of the *bhadrolok*.

An important ingredient in the new family was the nascent nationalism which was gaining ground among the *bhadrolok*.[6] The children would serve to end the misery of *Desh Mata*, not only from the religious and social shackles, but also mitigate the sufferings caused by a foreign rule. The family thus became the platform for national regeneration. Home management, bringing up the children, dietary habits and hygiene – all were important for the 'new family'. The *bhadrolok* lamented the prevailing condition of women:

> Had they possessed a bit of knowledge in their minds then they would have scarcely done all these abominable things, such as Ghatto Puja and Sashti Puja. But these things only occurred for want of education and therefore to release them from the heavy shackles of superstition, as I have greatly considered, is requisite in every respect and so it ought to be carefully implanted in every family.[7]

Naturally, mothers had a crucial role in this grand scheme of renewal and rejuvenation. As Pratapchandra Majumdar, one of the early authors on this subject remarked, "Because of the flaws of the mother, the child is ruined, when the child is ruined, the family is ruined, when family life crumbles, society decays; and when society is polluted, no nation can advance" (Bose 144-45). Atul Chandra Dutta, an early writer on domestic science wrote: "Well-trained children are the pride of the country; with bad training and corrupt morals, they only bring disgrace to the family and become (the) scum of the nation." The attempt to transform the family from within called for the dissemination of medical, educative, and relational norms.

---

[6] See Tanika Sarkar.

[7] Cf. Goutam Chattopadhyay.

While the sons were tasked with the grand project of 'nation building', the daughters were understandably excluded. A patriotic song clearly and unhesitatingly stated, "Who else but a son can rescue the *Desh Mata* from her sufferings!" (*Putra bhinna matri dainya ke kore mochon!)* Only the necessity for women's partial development was acknowledged to prepare them as future mothers and wives for an emerging nation. The necessity for complete development of a daughter as a person and as a human being remained unuttered and, in fact, was never thought of:

> The Women..., are for the most part totally ignorant of those moral qualities which are necessary... for being a useful member of society, particularly as society is at present constituted. We have hitherto considered the importance of female education only as connected with the proper education of man.[8]

Lilabati Mitra, an ex-student of Bethune School mentions that she felt 'happy' that education for girls had vastly improved by the early twentieth century – the girls were taking education seriously, they were obtaining bachelors and masters degrees and were teaching in colleges. She mentions in the same breath that the educated girls are making their homes havens of peace and pleasure and making their husbands, sons and brothers worthy of service to the motherland.[9] In the new scheme of things, the status of the child bearer was upgraded to that of 'mother' and 'educator' – provider of elementary education and 'medical auxiliary'. She was assigned a role that was greater than the one she had played previously; yet she remained subservient to man. Limited amounts of reasoning and knowledge were expected of women – enough for the upbringing of her children and management of the household but not more than that. The development of women was subject to the primacy of men in the society.

Whatever might be the avowed purpose, more girls were getting the privilege of education and exposure to the outer world and they used all available opportunities to prepare themselves as future mothers and wives of modern Bengal. Unconditional development, unfettered access to knowledge, strong as well as affectionate bonding with her own parents, and proud privilege of shouldering the responsibilities of her own family affairs certainly could produce a better woman and, more importantly, a better person. Such a woman could be a better mother with her own self empowered by rights and responsibilities. But that would be completely

---

[8] *BENGAL EARLY NINETEENTH CENTURY*, op.cit. 146-147.

[9] See Lilabati Mitra.

anachronistic to the existing patriarchal structure and mindset. Some even perceived that education would make women more committed to their traditional roles assigned within the family and society. An interesting letter to the editor of *Reformer* is very candid and ominous:

> Though envy is the quality of women in general yet it assumes a very prominent feature in the female society in India. Nothing but ignorance, the gross ignorance, the abominable ignorance, is the cause of all these mischievous consequences. I have witnessed myself, Mr. Editor, that it is impossible for any man to convince them of the truth of anything or persuade them to virtue and to make them obedient to the duties they owe to their husbands by virtue of sacred marriage and further, their godly reverence to gold creates frequent breach of peace between the wife and husband….thus the female of India, being deprived of all the advantages of education, become subject to the vices which render their lives unhappy forever.[10]

The mindset of the society can easily be assessed from the different responses received by three important girls' schools. Maharani Tapaswini and KD Carve founded girls' schools in Calcutta and Poona respectively and were educating young women from conservative homes to become better wives and mothers in a modern world. To gain the support of educated Hindoo families, they developed curricula dominated by home science and religious lore. Their syllabi were praised and got support from Hindoo gentlemen who believed that less emphasis on formal reading and writing and more on knowledge of sacred literature, myths and legends would be ideal for women to make them more suitable for the role of traditional women. They believed that the girls' schools run by Christian missionaries were 'demoralising' young Hindoo women and uprooting their social moorings. On the other hand, Pandita Ramabai, who founded girls' schools in Poona and Bombay, stood in direct contrast to the two educators mentioned above. Ramabai designed a remedial curriculum. Classes in psychology and botany were included to teach students about their own bodies and the physical world in which they lived. Industrial training was included to make them self-sufficient. Unlike the other two educators, Ramabai sought help from the Christian community. She had no faith in traditional Hindoo society which she thought created hindrance to the development of the democratic spirit. The traditional Hindoo society also did not support Ramabai's radical ideas and suspected her motives.[11]

---

[10] *BENGAL EARLY NINETEENTH CENTURY*, op.cit. 172-177.

[11] *The New Cambridge History of India Women*, op.cit. 32-63.

Kesab Chandra Sen, who breathed fire on the traditional Hindoo society, perceived women's education as necessary means to the end of providing sufficiently 'qualified' wives for the modern young men of Bengal and particularly discouraged teaching young women science or mathematics so as not to "unnecessarily tax the brain and distract from a girl's primary goal of happy, uncomplicated domesticity". On the other side of the issue was the *Brahmo* progressive fringe including Dwarakanath Ganguly – certainly one of the most fascinating modern Bengali reformers and early feminists whose proactive support helped Kadambini, the first female physician to overcome many obstacles.

Clearly, women, even those who were educated, were expected to keep to the straight and narrow path defined by society – education was supposed to help them perform their newly-interpreted traditional role better. Even for her new role, her development as a person – self reliance, self confidence, self realisation and emancipation – was not part of the agenda or the goals of her education. Although it will be unfair to judge the past in the light of today's thinking, we cannot help observing that on the issue of women's education, the most radical of reformers and rebels were as traditional as anybody else. Very few could de-link women's education from tradition and perceive it as a goal unto itself.

Therefore, modern women of the nineteenth century were born with an unspoken understanding. They knew that the purpose of their education was to become worthy wives and better mothers of a new generation of *bhadroloks* and to bring up their children by providing them rudimentary education and (knowledge of) hygiene and rid herself of taboos and superstitions. But she was not expected to go to the extent of questioning her subordinate role in the family or her deprivation of parental affection and subsequent uprooting from the family at a tender age and transfer to an unknown family.[12] Education was imparted to girls with the aim of making a new self which would maintain a perfect blend of the spirit of enquiry and the passion of submission to appropriate norms of traditional women (Roy 153). This idiosyncratic demand of this emerging society is very intelligently pointed out by the eminent writer, Lila Majumdar, who asserted that women are very intelligent and that's why they can pretend to be dumb for such a long time.[13]

Expanding female education, even with this very limited purpose and targetting a very small section of society, proved to be a very difficult task both for the colonisers as well as the new *bhadroloks*. The main hindrance was the very customs which the education was supposed to reform.

---

[12] See Malavika Karlekar.

[13] Lila Majumdar. *KherorKhata.* Kolkata, Anondo.

According to tradition and customs, a girl was a responsibility and a liability for her parents and she had to be married off before she attained puberty. Therefore, it was difficult to find girl students beyond the age of ten. As temporary custodians, parents did not feel the urgency of educating a girl child. For them, it was more important and safer to place the girl in her in-laws' house than to send her to school. Dearth of lady teachers was another major problem. Parents were not ready to accept male teachers for girls. An interesting feature emerged from government records. Educated and established Hindoo gentlemen of Dacca were very eager to educate their wives. But the parents were not confident of sending their daughters to schools. One school for adult women was running very well, which was an exception, attesting to husbands' desire to educate their wives while day schools for girls flopped. Government considered closing these schools:

> The girls' schools in Dacca Division are in such a very unsatisfactory (to me) condition, that I should feel no regret were government (with two or three exceptions) to abolish all at a blow. The children are, on the average, very young; in a good girls' school, there may be one girl of eleven, three of eight to nine, and a dozen of five to seven years of age. Neither the parents nor the teachers feel any interest in the school beyond the Government grant, the school sits an hour or so per diem, and that very irregularly, and the standard attained bears no approximation whatever to that to attained by boys of the same years. The present system indeed while effecting little real good involves a tacit acknowledgement by Government that keeping girls after twelve years old (under) lock and key is a system to be maintained.[14]

Performance of Dacca Female School during 7 years (1864-1871)[15]

Salary at which appointed	Number of mistresses	Christians	Brahmins	Kahetrias	Kaests	Bairaginees	Number of mistresses sent to village	Number of mistresses sent to town	Remarks
Rs 5	1	1					1		
Rs 10	4	4						4	
Rs12	1					1	1		
Rs20	4	3				1	1	3	
Rs 25	2	1				1	1	1	
Rs 30	5	1	2	1	1		4	1	
Total	17	10	2	1	1	3	8	9	

The reluctance and lack of sincerity in imparting education to girls in Dacca are very explicitly expressed in the above-mentioned government

[14] General Education: March 1872, Proceeding no.94, WBSA.
[15] Ibid.

records. The approach towards girls' education was not very different in Calcutta either. Lilabati Mitra, a student of Bethune School reminisces:

> The girls came to school by nine in the morning and played till eleven. The school started at eleven. The teachers paid very little attention (to) what the girls were learning. I kept the books under a basket after returning home ... The teachers were careless about covering the topics...Some evaluation was done but the failed students were allowed to study in the next class.

Performance of the boys' schools during the same period was much better – both the students and the teachers were much more sincere. That English education was necessary for getting jobs and to deal with the new regime acted as an obvious motivation for the boys. The importance of educating girls for their own sake was out of question. Girls' education, even for the sake of the family and society, was still not part of popular imagination. Thus the involvement of institutions through which knowledge was transmitted here varied along the gender line.[16]

Students of Bethune School and their fathers' socio-economic backgrounds[17]

NAMES OF PRESENT PUPILS	PRESENT AGE	NAMES OF PARENTS OR GUARDIANS	PROFESSION
Shookada	11	Hurrodabe Chatterjee	Formerly Sheristadar of Magistrate of Zillah Court. At present out of employment.
1st Nistarenee	9	Do	Do
1st Sukee	6	Do	Do
Beruda	9	Ramchunder Ghosal	Sheristadar of Stamp Office

[16] The involvement of specific institutions through which knowledge was being transmitted played crucial roles in the process of transmission. See *Texts of Power Emerging Disciplines in Colonial Bengal*, ed.Partha Chatterjee, op.cit. "The various modern knowledges arrive in the colony in the nineteenth or early twentieth centuries already formed as disciplines, their objects and boundaries defined, their conceptual apparatus in place, and their authorities firmly established (in the metropolis of course). The only interesting question here seems to be the institutional one, that is, of the specific educational, literary, cultural, or professional institutions through which the modern disciplines are transmitted in the colony and the nature of their appropriation within various non discursive practices." (P.1-29)

[17] General Education, 16th March 1854, Consultation no 133,WBSA.

Gyanoda	8	Taranauth Turkobachosputty	One of the Pundits of Sanskrit College
2nd Nistarenee	9	Russick Lall Sen	One of the English teachers at Juggernauth Pooree
Khettore	10	Cossy Nauth Nundy	Merchant
Netrokally	8	Do	Do
Koosum	6	Do	Do
Rajessurree	8	Hullodhur Auddy	In the employ of Strand Mill & Co
Koosumcoomaree	5	Shib Chunder Dutt	Writer
Thakoo Dassee	8	Gooroo Churn Paul	One of the Writers of the General Treasury
Jaunoby	5	Do	Do
Sowdameny	8	Debendranath Tagore	Merchant and zemindar
Bamasoonderie	10	Mohes Chunder Chatterjee	Hindoo Priest
Raj Lukee	10	Shib Chunder Dey	A Banker
1st Dube Ranee	9	Nundolaul Paul	Merchant
2nd Lukee	6	Gour Mohun Paul	Ditto
Kameney	6	Gobind C Seal	Banian of Calcutta Docking Company
1st Radaranney	6	Nobokoomar Paul	Cotton Merchant
Soorut Koomary	6	Nilcomul Mitter	Banian
3rd Lukee	4	Kishoree Mohun Paul	Ironmongery shopkeeper
Moheeny	6	Rajender Nauth Bose	Writer
Jadobee	4	Gobind Chunder Seal	Native Doctor
Kooronee Dussee	5	Ramkony Coondoo	Cotton Merchant
Toolkooaree	4	Shama Churn Sen	Head Writer of Police at Chief Magistrate's Office
2nd Dube Ranee	8	Kheter M Paul	Thread Merchant
Ranee	5	Shama Churn Paul	Cotton Merchant at Moorshedabad
Rajranee	6	Ram C Paul	Cotton Do at ditto
3rd Dube Ranee	6	Dwarkanauth Sreemolee	Ditto at Tumlook
Unnoo Muttee	6	Nundo Coomar Set	Has no situation
Surosotty	5	Radhanauth Coondoo	Banker
Needoomonee	9	Cossinauth Paul	Has no situation
3rd Nistarenee	5	Chunder K Sett	Banker
4th Lukee	7	Modoosooden Sreemelee	Thread Merchant
4th Nistarenee	9	Russicklaul Paul	Cloth Merchant
Prasad Dossee	5	Nokoor Chunder Dutt	Banian

4th Dube Ranee	6	Sib C Sreemanik	Cloth Merchant
5th Nistarenee	5	Bannymadhub Dhur	Writer
Burcoohannoo	5	Sree Nauth Chatterjee	Hindoo Priest

Although the imperialists did not have much respect for the natives' culture and customs and would use every opportunity to denounce them, their zeal for social reform through legislation and coercion suffered after the 'sepoy mutiny'. The government became very circumspect. Instead of imposing it from outside, the government wanted the effort for radical social reforms like girls' education to come from within the indigenous community – otherwise the efforts of the government would be fraught with risks. There was apprehension that girls' education was bound to create upheavals in society. A report – insightful and prophetic enough to be from a top-ranking bureaucrat of foreign origin – notes:

> "The Government of India need not say that it recognises to the fullest extent the fact that the encouragement of female education is an object, the importance of which cannot be estimated too highly. But it fears that if the greatest care and caution be not exercised, the efforts of the Government to promote this object may tend rather to frustrate it, and to stimulate the opposition of the ignorant and suspicious. The Governor General in Council is far from satisfied that (except in the immediate neighbourhood of the Presidency Towns) the native community has yet really shown any spontaneous desire for the extension of female education; indeed there is ground to fear that the action already taken in that direction on the part of Government has, in some places, been regarded with mistrust; nor is it surprising that this should be the case, it must be obvious, even to the most ignorant among them, that the natural result of the general extension of female education would be to place the domestic relations of every family on a new footing and to break up existing social habits and traditions. Even when these results themselves are fully accepted as beneficial, the interference of foreign Rulers to affect them will probably be distasteful. Far more must this be the case when such changes are opposed to widely prevailing customs, or to deeply rooted and long established prejudices. It is on this ground that His Excellency in Council considers that if measures for female education be set on foot by external influences, and specially by that of Government, the Native community is not likely actively to co-operate in forwarding them, but on the contrary, will receive them with apathy, if not with opposition.
>
> It is only where large experience has already taught the people the advantage of education generally, and the special benefit of the changes which the spread of female education would effect, that they can reasonably be expected to feel the want of means for female education, and it is only then that any demand for the supply of these is likely to arise; but it is probably only to a very limited extent in the Presidency Towns, and

> among those classes who have participated in the advantages of superior English education, that such a want has yet been felt."[18]

Female education and the custom of early marriage were antithetical to each other – no compromise or working around was possible. Only in case of the *Brahmo* girls, it was comparatively easy to pursue school education up to a certain age, typically fifteen or sixteen. For Hindoo girls, early marriage and the apparent sanction accorded to it by the scriptures, made any progress very difficult, if not impossible. The Hindu Widow Remarriage Act, 1855, Civil Marriage act 1872, made little impact on Hindoo society.[19] Even Iswar Chandra Vidyasagar, the great scholar and social reformer could not oppose early marriage of girls in spite of wholeheartedly supporting Drinkwater Bethune[20] in his endeavour to educate native girls. While giving his opinion on the Age of Consent Bill, he said:

> I am not able to give unqualified support to the Bill. If passed into law, it will prevent the performance of the ceremony of Garbhadhana in all cases where wives attain puberty before they are 12 years of age. Garbhadhana is a ceremony which is obligatory and is generally observed in Bengal. It has to be performed by the husband on the first appearance of the menses in the wife. It is not necessary to quote many texts in support of this position. One text from Parasara, who is the special authority for this age, i.e. Kali Yuga will suffice, 'He who, being at hand, does not approach his wife after her bath on the occasion of the menses is certainly guilty of the terrible sin of destroying the foetus.' As some girls may menstruate before they complete the age of 12, the ceremony cannot be performed in their case if the Bill is passed and people will be justified in complaining of a legislative prohibition of a compulsory religious rite.[21]

The *Brahmo Samaj* collectively framed a rule that girls should not be married off before the age of fifteen. But when it had to be implemented at the individual level, not everyone could rise beyond the age old custom. Keshab Chandra Sen, a leading member of the *Samaj* married his daughter off at the age of fourteen. Even Rabindra Nath Tagore was not exemplary in this regard.

Thus, women in the nineteenth century took the first tentative steps towards modernity. The journey of nineteenth century women towards

---

[18] General Education, July 1868, Proceeding no.76, WBSA.

[19] See Sunil Sen.

[20] See Jasodhara Bagchi.

[21] "Note on the Bill to amend the Indian Penal Code of Criminal Procedure, 1882", Bengal Judicial Proceeding, June 1893,WBSA.

modernity started on the very basis of patriarchy. That women should be acknowledged and groomed as complete persons was unthinkable at that time. In spite of the limited scope and narrow course defined by patriarchy, women's education and their endeavour for self-development gathered momentum. In course of time, the emerging society expected women to perform many more tasks and this opened up opportunities before them. In some cases, the conservative attitude of society advanced the cause of modernity. Particularly in the spheres of medicine and education, the necessity of qualified native women was intensively felt.

Since most women would not visit a male doctor or even a lady doctor of a different religion, the need for native lady doctors was acutely felt. Thus, to make modern treatment accessible to women, the world of medicine opened up its doors to Indian females. In a letter, dated 4th June 1883, Mr G Ballet, who was then officiating as Director of Public Instruction, submitted a proposal for the admission of females to the Medical College in Calcutta. In 1882, the Director of Public Instruction urged Medical College authorities on the advisability of admission of females to the college on the ground of great alleviation of suffering which, it was thought, would result if there was a body of qualified practitioners to whose admission to *zenanas* there would be no objection.[22] Even in the medical colleges, the co-education system was well accepted.[23]

But the opportunities and facilities were opening up strictly based on society's need. Even a single step outside the said requirement was questioned. In the 1880s, women were entering Calcutta Medical College and co-education was accepted. In 1897, a debate raged in the corridors of educational administration when two *Brahmo* girls whose fathers were in high positions (Dr PK Roy, Professor of Presidency College and Mr RN Roy, Deputy Comptroller and Auditor General) were admitted in Presidency College. That the acceptance of co-education in medical

---

[22] In 1883, the then D.P.I. Mr. G. Bellet stated "I have constant experience that it is too often the case that many indeed the vast majority would rather die than submit to examination by a male practitioner." See File 88, Proceeding no 1-8, General Education July 1883.

[23] "On the question of mixed classes, there is much more to be said and I fully understand the objections taken to them. I think, however, that these objections are exaggerated, and that women who have so far conquered the customary ideas (I will not say prejudices) as to what is modest and feminine, as to be ready to study medicine in company with men may be safely considered able to take care of their own morals, while the fact that their every act is liable to misinterpretation." [File 88, Proceeding no.1-8, July 1883, op.cit.]

college was based on society's need was clarified in one government correspondence,

> The necessity for admitting female students to the Medical College arose from the fact that there was no separate college for imparting medical instruction to the females. It may be mentioned that as the Zenana Hospitals which was then under consideration required that the Zenana Hospitals should be officered chiefly by lady doctors, the admission of ladies into the Medical College to qualify themselves for such offices by receiving proper training in that College was one of the grounds for allowing them to study in it. But it is for consideration whether the ladies, who will take up the general course of studies, viz, FA, BA and MA, as two ladies admitted to the Presidency College have taken, cannot attain their object by studying in the Bethune College Calcutta, where such classes exist, and which is intended for the exclusive use of ladies.[24]

Not only the administration, but all leading newspapers had "made some remarks about the admission of the two young girls into the Presidency college."[25] The College authorities had conversations with the fathers of these two girls who stated that "their desire was to give them the very best education that was possible and that could not be obtained in the Bethune College."[26] The desire of the parents of the girls came under severe scrutiny. The young ladies came from *Brahmo* families. *The Cosmopolitan and Liberal*, a journal which represented the *Brahmo* cause, did not approve of their admission to the Presidency College which was a boys' college till then:

> Considering that there is already in Calcutta such a well-managed institution conducting exclusively for the education of young women it is decidedly a retrograde measure. While the Universities of Oxford and Cambridge have shown themselves peculiarly orthodox in matters of female education, it is to be deeply regretted that some unwitting advocates of radical reforms in India have rushed in where angels dare not tread.[27]

The mandarins of progress were very cautious to ensure that women did not cross the limits they set for them. The Victorian doctrine of 'separate spheres' located domesticity and the home as women's realm where she would remain protected from all dangers and temptations.

---

[24] General Education File B, File 4c/20, Proceeding. 3, September, 1897.

[25] "Admission of Girls into the Presidency College", To the Editor of Indian Mirror, 5th August, 1897.

[26] General Education File B, File 4c/20, op.cit.

[27] Ibid.

Marriage and motherhood were identified as the central goals for them. To discourage women from higher studies, it was claimed that medical studies had clearly indicated the harmful effects of excessive study on the menstrual cycles of women and their ability to conceive (Sen 2-3). In 1885, a Government report of Bengal stated:

> In England a limited number of girls 'study' in College. The mass of the girls' education of the middle classes is paid for, in order that the girls may learn 1) music, 2) drawing, 3) French (or German). All the English boarding schools teach accomplishment.[28]

The ladies whose husbands were appointed in senior administrative positions in India were groomed so that they could perform the role of wives of imperial rulers to perfection.

It seems interesting that in some respects, colonies showed more significant advancement than England. In England, no woman student received any degree from Oxford University before 1920, whereas in 1884, Chandramukhi Bose received her MA degree from Calcutta University. In USA, co-education system was introduced in most of the colleges by the end of the nineteenth century whereas in England co-education was not introduced other than in medical colleges. In this given context, the admission of two girls to Presidency College may be seen as an exemplary step towards modernity. The two girls' parents, who were products of Western education, thought, dreamt, and aspired in new ways that were different from the so called 'enlightened 'Western world. It may be mentioned that apart from the above-mentioned two girls, another girl named Saralabala Biswas of Cuttack was also admitted at the same time to Ravenshaw College of Cuttack under the Director's order. It seems interesting that the colonial rulers also took the liberty of using their own discretionary powers ignoring the stiff gender and class-based dictums of their parent country. In this case, the liberal decision was finally scrutinised by higher authorities. To give an explanation for such liberal decisions, Dr Martin said:

> I never contemplated that any opposition would come from the fathers of the male students of the college and from what I learn from Mr Rows, everything is gong on in a most satisfactory manner. These ladies are properly escorted to college every day and when they go from one class room to another, they do so with one Junior Professor in attendance.[29]

---

[28] Education Commission, Proceeding B 5/6, File 13, January 1885,WBSA.

[29] General,Education File B, File 4C/20 op.cit.

The contours of social change were not very even within the colony either. Calcutta and Cuttack produced remarkably progressive parents who aspired for their daughters' development surpassing the colonial as well as contemporary British society's requirement but the scenario of villages and mofussil towns was not so promising in this respect.

Data on women professionals in mofussil schools[30]

Name of the Mistresses	Dt of appointment	Pay per mensem	Name of the school to which appointed
Martha	1st may 1864	Rs 10	Dacca Female Normal School
Sukhini	22nd June 165	Rs 5	Mushigunge girls' school
Bisaka	6th June 1865	Rs 12	Mulkhanogur girls' school
Radhamani Devi	29th June 1866	Rs30	Sherepur school
Phulmani	10th April 1867	Rs 10	Dacca Zenanas
Basakha	30th April 1867	Rs 25	Rajshahye Balihor school
Haripara	3oth April 1867	Rs 20	Comilla girls school
Isvari	20th May 1867	Rs10	Dacca Zenana School
Radhurnani Devi	19th October 1867	Rs 30	Rajshahyee Female Normal School
Parawmani	5th September 1867	Rs 20	Noakhalli girls school
Bhagavati	7th December 1867	Rs 20	Faridpur Girls School
Pabitra	10th July 1869	Rs 10	Dacca Female School
Janaki Devi	14th September 1869	Rs 30	Rungpur Goalpur School
Saressvati	14th March 1870	Rs 30	Dinajpur Ranigunge School
Hara	14th June 1870	Rs 20	Suhajatpu School Serajgunj
Daya Mai	29th Decdember 1870	Rs 30	Rajshye Dighaputya
Bamasundari	24th April 1871	Rs 25	Dinajpur School

30 General Education March 1872, Proceeding 101,WBSA.

In the 1860s, girls' schools were established to impart basic knowledge on reading and writing to the girls. These schools also served as training grounds for female teachers. The students who enrolled in school, after completion of their course, had to be in service as lady teachers in schools in different districts. The proportion of students of higher cast Hindoo families was very small, whereas the presence of converted Christian and *bairaginee* sects was prominent. The government expressed its distinct inhibition in admitting *bairaginee* students. The *bairaginees*, whose 'moral' status suffered from ambiguity, could not be accommodated in the grand planning of preparing 'educated *bhadromahilas*'. Rather, the government anticipated that their presence would create further hindrance to participation from higher cast Hindoos in female education. The Deputy Inspector of Schools of Bikramore recommended:

With a view to raising the school in the estimation of the people and to hold out inducements to women of respectable families taking their admission into it, I beg leave to suggest as follows:

1. That the *byraginees* be at once removed from the school and that their places be supplied by good moral character.
2. That provisions be made for board and lodging for the students, who have no means of comfortably residing in town, within the premises of the school.
3. That a committee of respectable European ladies be formed for the inspection and management of the school.
4. That rules be made prescribing the term for which the students are to continue in the school.
5. That no student be admitted to the school without the production of a certificate from the Deputy Inspector of Schools or other respectable gentlemen testifying to their good moral character; in the case of native Christians, certificates from missionaries would be most desirable.
6. That the deputy inspectors be instructed to send, if possible, two or more students from their respective districts, as it would tend to supply mistresses for their districts at a comparatively low pay.[31]

Thus, the process of educating women of 'respectable families' in Bengal was not at all very smooth. Barring a few enlightened urban families (especially *Brahmos*), the major part of the population did not respond very positively. The headmaster of Dacca Normal School, Shri

[31] General Education, March 1872, Proceeding no.97,WBSA.

Mahesh Ganguly, was not very hopeful about the future of women's education:

> A school of a higher social class is, under the present circumstances, an impossibility as has, I think, been clearly evinced by the failure of the adult female school... would not have come to the miserable end it did, *perdanishanees* going to school is a contradiction in terms, and must be (as it has so long been, and I do not know for how much longer it may be) greatly out of countenance in the higher social class.[32]

However, the inner resistance of higher social classed to educating their females decreased with time. In the late nineteenth century, in England as well as in colonial India, public space was slowly opening up for women. But the perceived limitations of women's participation in public space were constantly highlighted in different discussions. It was said that, "Not one girl in a hundred would be able to work up the subjects required for the Indian Civil Service Examination in the way which boys do."[33] In Bengal, when women got entry to the field of medical education, their perceived limited utility in this field was pointed out by one administrator: "If there were a supply of native lady medical practitioners in Bengal now, I think they would only be useful as obstetric practitioners. For general diseases, and in difficult obstetric cases, the native community, I am of opinion, would prefer male attendants."[34]

After receiving higher degrees from Glasgow and Edinburgh, Kadambini Ganguli returned to Calcutta and started her practice. This level of success in the career of a woman was beyond the imagination of native society, especially women. When Kadambini went to visit female patients, the women of the locality addressed her as *daima* (untrained midwife).

In the case of the admission of two lady students to Presidency College, the impossibility and absurdity of the access of women in every sphere of public life was specifically mentioned. The two lady students aspired for a better science education and got admission to Presidency College; yet they were constantly reminded of their limited access to the professional world:

> Recently, Miss Cohen applied to the High Court for enrolment as an Attorney but her application was refused by the court. So, whatever high university degrees the native ladies may possess, they have no chance of

---

[32] General Education March 1872, Proceeding no. 100, WBSA.

[33] *Women and Empire*, op.cit.

[34] General Education, File 88, Proceeding no.1-8, opcit.

> becoming pleaders or holding responsible judicial and executive offices under Government. The utmost they can get is some educational appointments under this Government for which they can still qualify themselves with the education they may receive in the Bethune College.[35]

Although women were being encouraged, at least in some sections of society, to get education and 'intrude' into certain arenas inaccessible to them till then, the regulators of society and the administration were always looking over their shoulder in a bid to ensure that women did not take 'undue advantage' and in the 'heady rush for modernity', women did not abandon their 'traditional roles' as homemakers and upholders of social values and religiosity. Even the 'enlightened' women felt obliged to be equally adept at the duties of daughters, wives and mothers and maintaining the sanctity of the home and hearth. In more modern times, the communists, who wanted to end exploitation and inequalities in the society, were blissfully unaware of the inequalities they inflicted on women comrades. Chakilam Lalithamma, a woman activist of the Telengana movement, reminisced that male comrades would not participate in cooking and serving in the communes (Chakraborty 86-87). Those who rejected religion and religious customs in other spheres of life would not take the same to its logical conclusion in the case of women. Even the revolutionaries were traditionalists in the realm of women's liberation.

Works Cited

Bagchi Jasodhara, "Socialising the Girl Child in Colonial Bengal". *Economic and Political Weekly.* 28. 41 (Oct. 9, 1993): 2214-2219. Print.

Bose, Pradip Kumar, "Sons of the Nation: Child Rearing in the New Family". *Texts of Power Emerging Disciplines in Colonial Bengal.* Ed Partha Chatterjee. Minneapolis: U of Minnesota P, 1995. 144-45.

Chatterjee, Partha ed, *Texts of Power Emerging Disciplines in Colonial Bengal.* Kolkata: Samya, 1996. Print.

Chattopadhyay, Goutam, ed., *Bengal Early Nineteenth Century Selected Documents.* Research India Publications, June 1978. 128-130. Print.

Chakraborty, Ishita, Paine Barnali & Bandyopadhyay Krishna eds., "Naishabda Bhenge, Atmakathane Bharatiya Nari". Kolkata: Khonj Ekhan Parishad and Stree, 2005. 86-87. Print.

[35] General Education, File 4C/2, Proceeding no. B20,op.cit.

Forbes, Geraldine. *The New Cambridge History of India: Women in Modern India*.Delhi: CUP, 1996. Print.

Karlekar, Malavika. "Women's Nature and Access to Education". *Socialisation, Education and Women: Explorations in Gender Identity.* Ed. Karuna Chanda. New Delhi, 1988. 133. Print.

Mitra, Lilabati. "Puratan Bethune School" (Bethune School of the past). *Sekeler Katha* (*Tales of the Yore*). Eds. Abhijit Sen & Abhijit Bhattacharjya. Kolkata:Naya Udyog, 1997. Print.

Roy, Raka. "Conformity and Rebellion". *From the Seams of History: Essays on Indian Women*. Ed. Bharati Ray. Delhi, OUP, 1995.153. Print.

Sarkar, Tanika. "Hindu Conjugality and Nationalism in the Late Nineteenth Century Bengal", in *Indian women: Myth and Reality.* Ed. Jasodhara Bagchi. National Seminal Papers, 1989. Print.

Sen, Indrani. *Women and Empire, Representations in the Writings of British India 1858-1900*. Delhi: Orient Black Swan, 2002. 2-3. Print.

Sen, Sunil. *The Working Women and Popular Movements in Bengal.* Calcutta: K.P. Bagchi, 1985. 24-25. Print.

CHAPTER SEVEN

WAS TAGORE A FEMINIST? RE-EVALUATING SELECTED FICTION AND THEIR FILM ADAPTATIONS

SOMDATTA MANDAL

Let me begin by stating that this paper has nothing exceptionally new to offer. Like most scholars enamoured of Tagore's polyvalent personality, especially his treatment of women characters in his novels and short stories, I ask the simple question: Was Tagore a feminist? And like all readers, I come to the conclusion that indeed he was one. But since he was much ahead of his time, he was neither accepted wholeheartedly as one by the intelligentsia of the times, nor could he sustain radical views in his female protagonists till the end in many of his novels and short stories. This resulted in compromises, in ambivalence and ultimately, in the dilution of strong feminist viewpoints. This paper will focus on Tagore's stories that profess strong feminist values and also their representations in certain cinematic adaptations. But before analysing this point, it would be worthwhile to briefly reiterate the basic tenets of feminism.

Feminism is both a political stance and a theory that focuses on gender as a subject of analysis when reading cultural practices and as a platform to demand equality, rights and justice[1]. Feminism's key assumption is that gender roles like 'daughter' or 'mother' are not natural but social because a woman has to be trained to think, talk and act in particular ways that suit the role. Feminism's key political and theoretical stance is this: The inequalities that exist between men and women are not natural but social and not pre-ordained but created by men so that they retain power. Religion, the family, education, the arts and knowledge systems are all social and cultural 'structures' that enable the perpetual reinforcement of this inequality.

[1] For detailed analysis of different facets of feminism see Pramod Nayar, 83-84.

When Tagore started writing creative fiction, the West was undergoing the first wave of feminism (1830-1920) with its suffragette and civil rights movements. Educated Indians, including Tagore, were not unaware of the New Woman concept of the West. In India, the slow but unstoppable progress made in the field of women's education, the emergence of a small but not insignificant band of woman writers and the increasing participation of women in the struggle for freedom further led to the increasing awareness that a man could perhaps no longer dominate his wife to the extent he had so far taken for granted. The desire for autonomy inevitably began to percolate down to the level of familial relationships. In Tagore's own household, the women of Thakurbari (as his ancestral residence is commonly known) were all educated; many of them were creative writers; many of them adopted Western lifestyles selectively, and were known for their excellence in the performing arts. However, the women of Tagore's household generally conformed to traditional norms regarding marriage and fitted themselves into the expected roles of homemaker and caregiver as wives and daughters. This binary state of being a free woman but, at the same time, conforming to socially set stereotypical roles often resulted in a dichotomy which we find being adopted by Tagore in delineating many female protagonists of his novels and stories. The centuries-old conservatism in society which dies hard against change is attacked in his writings.

I. The Woman Question

We are all aware of the fact that Tagore made extensive experiments with women's emancipation and sexual freedom in many of his novels and short stories in order to express his views on the gender question. But the trend of his feminist ideas is not consistent or uniform. It moves like a graph going up and down – feminist ideas at the zenith and conservatism/traditionalism at the nadir. Writing at a time when patriarchal society was not liberal enough to concede its women any or any attempt at progress and giving equal rights to women shook society to its foundation, Tagore was not only naturally sensitised to the plight of women in Indian society but actually wanted to establish the individuality of women because it was individuality that ensured their dignity, self-respect and sexual independence.[2]

---

[2] See Santosh Chakrabarti, *Studies in Tagore: Critical Essays* pp.60-61 for further details.

The plight of women living in a marginalised state in a male-dominated patriarchal society is found in his early short stories like "Denapaona" which shows the travails and tribulations of a girl who had not brought sufficient dowry to her husband's home. The woman character in this story is very vocal. In a subsequent story called "Haimanti", the girl of the same name dies a slow death after making a firm protest. The dowry issue is again the focus of "Aparichita", a story published in the same year. In several other short stories like "Poyla Nombor", "Manbhanjan", "Bostami" and "Strir Patra", Tagore creates the space for an unusual female protagonist – a strong-willed, sensitive, moral and independent woman who can challenge/question dominant patriarchal ideology and pose an alternate notion of female subjecthood outside its prescriptive norm. In "Poyla Nombor", Anila finally leaves her husband who, engrossed in his world of intellectual pursuits and male bonhomie, had taken her for granted. Her brief note to him – "I am leaving. Please do not try to look for me. You shall not find me even if you try" – comes to him as a shock and revelation.[3] Far more intriguing (both for the narrator and readers) is the knowledge that she has left an identical note for Sitangshu Moulik, the man who admired and worshipped her from a distance and was even willing to lay down his life for her.

The idea of a woman's right to choose her own path is elaborately illustrated in "Streer Patra".[4] In this story, Tagore questioned the system of arranged marriage and the entrapment and enslavement of women as wives. Though their husbands provide them with shelter, security and sustenance, the three women portrayed in the story, Mrinal, Bara Bou and Bindu, have loveless marriages. But the obvious agenda of "Streer Patra" is eliminating the role model of the beautiful and submissive wife, Mejo Bou, and registering the birth of Mrinal. This metamorphosis from being a non-entity to a person conscious of her identity, of course, does not happen all on a sudden. Tagore constructs a very sensitive narrative in which he systematically records Mrinal's responses till she resolves not to return to her husband's house which never became a home for her in the fifteen years she lived there:

---

[3] See *Galpaguccha III*, p. 624.

[4] In her article "Breaking Free: Is Rabindranath's 'Streer Patra' A Feminist Text?" Sanjukta Dasgupta elaborates on this issue and several ideas and quotes for the discussion of this text are borrowed from her.

> But I shall not return to 27, Makhan Baral Lane ever again. I saw what happened to Bindu. I have realised the position women have in this society. I don't want to go through it anymore.[5]

What is significant is the fact that having been married at the age of twelve, Mrinal's rejection comes through a letter – written by a literate woman – beginning with the reverential *Sricharankamaleshu* (Your revered lotus feet) and ending with the supreme self assurance *Tomar charantalashroy chhinno* (Free from the shelter of your feet) – signed not as 'Mejo Bou' but as 'Mrinal'. It is obvious that her self-discovery is complete.

Several critics believe that the journal *Sabujpatra* was really the watershed in Tagore's concept of women's empowerment. We are all aware how "Streer Patra" invited protest letters from the conservative segment of the public when it was published in the Sravan issue of *Sabujpatra* (1321BS/1914). Bipin Chandra Pal's "Mrinaler Patra", Lalit Chandra Bandyopadhyay's "Swamir Patra" and his anonymously written "Streer Prokito Patra", published in the monthly journal *Narayan*, were not only scathingly critical of Tagore but predicted chaos and doom in family and social life. According to critic Niharranjan Ray, this progressive and feminist outlook of Tagore's, espousing women's emancipation, was not only the result of his personal thinking but was also imbibed from the progressive thinking of the West. He observes:

> The value of the woman is a comparatively recent discovery, the result of socio-economic evolution. This discovery was first made in Europe and other Western countries at the beginning of the 19th century… but this discovery made its impact in the third and fourth decades of that century and its full-fledged expression was observed after the First World War. "Streer Patra" was the first indication of the impact of that wave on our calm, unruffled shores (375)

Tagore had himself admitted that this short story was his first attempt at writing a 'pro-woman' text and in a letter written on May 17, 1941, he mentioned, "It's in my short story "Streer Patra" that I supported the woman's cause for the first time.[6]" Scholars and critics have often cited the similarity between Mrinal and Nora's predicament in Ibsen's *A Doll's House.* According to Maitreyi Chatterjee, initially, Tagore, at a young age, seemed uncomfortable with Nora's decision. But at the age of fifty-three,

---

[5] "Streer Patra", *Rabindra Rachanabali* VII. Kolkata: Visva-Bharati, 647.

[6] See Tanika Sarkar, "Mrinal Onno Itihasher Sakkhor". *Desh,* August 5, 2000: 35.

he reinterpreted Nora as Mrinal in the context of a daughter-in-law in an urban joint family in Calcutta.[7]

In this context, it is worthwhile to mention that in 1927, a little over a decade after the publication of "Streer Patra", Sushama, described as 'a niece of Tagore's' toured America and delivered lectures. She was a married woman, a mother of seven children, and had a barrister husband. However, though she could speak English and was aware of colonial culture, her personal responses regarding the position of women in India were conventional and conservative:

> The supreme, traditional virtues of the Hindu woman are fidelity, sincerity and self-sacrificing love. A wife subordinates her wish to those of her husband.[8]

Also, in a lecture hall in New York, she spoke about her pride in Hindu culture, marriage and marital relationships:

> Your idea of marriage, companionate marriage and love seems very strange to us. Your divorces startle us. We believe in the holiness of marriage, consider it a sacred and divine union of two souls. Our marriages are regarded as permanent, separation or divorce unspeakable, we stay married.[9]

As mentioned earlier, Tagore's portrayal of the emancipated woman was not uniform in all his stories and novels. In *Chokher Bali* (1903), his feminist ideas take a backseat to some extent and if we read his

---

[7] In an article in *The Sunday Statesman*, 4 August, 2002, Chatterjee states:
"But Ibsen's play received hostile notices because it encouraged women to question the society about home and family. Even Tagore was annoyed with the ending of *A Doll's House.* He felt Nora's rejections were in excess of her husband's condemnation of her. He felt Nora's rejections were in excess of her husband's condemnation of her. He had even gone on record saying that in the Indian family structure, women brought up on the ideals of patience and self-sacrifice were more appreciated and respected, and that Nora's situation was unthinkable in India. But that was Tagore in his young days. When he was wiser, more thoughtful and less biased at the age of 53, he made his heroine Mrinal in "Streer Patra" (1914) reject home and family in search of a new meaning in her life as a human being. Nora was reinterpreted in the context of a daughter-in-law in an urban joint family residing at 27, Makhan Baral Lane."

[8] Sanjukta Dasgupta translated the quotes that were originally published in Bengali. See Chitra Deb, *Thakurbarir Andarmahal.* Kolkata: Ananda Publishers Pvt. Ltd, 1983: 153.

[9] *Ibid.*

introduction, it becomes obvious that he had succumbed to societal pressures to present the ending to the novel that he did. Though the personality and sexual independence of Binodini were the inevitable results of twentieth century concepts of individualism, in the novel, Tagore had also portrayed two contradictory personae of Binodini as a woman – she is the source of familial and societal good/well-being and she is destructive. However, Tagore did not lower the Mahendra-Binodini relationship to the level of illicit sexuality and, on the other hand, kept the Behari-Binodini relationship at the idealised platonic level.

Unlike *Chokher Bali*, Tagore was very progressive in portraying the triangular love of Bhupati, Charu and Amal in his story "Nastaneer" composed in the same time span. Again, in *Chaturanga* (1322 BS/1915), he sought to strike a hard blow to the citadel of conservatism by bringing about widow remarriage. Damini, the young widow in this novel, is a rebel after the death of her husband and the desires of her body and mind express themselves in rebellion. In this novel, Tagore has not only made a bold and decisive assault on age-old conservative thinking by bringing about a marriage between Sribilas and the widowed Damini, but he also gave due importance to the individuality of a woman by prioritising Sachish as the object of true love over Sribilas in the heart of Damini. That Damini rises above Binodini is proved by the fact that the question of society and public protest do not deter her from the path of marriage.

In *Ghare Baire* (1916), Tagore turns a traditionalist again. Bimala wants to live the life of a traditional Hindu woman according to age-old Indian customs. But Nikhilesh's expectations are of a strange and different kind. By taking her out into 'the world'/ 'outside' and allowing her sufficient independence, he wants to prove his own superiority and claim greater love from her. That she is thrown off the domestic fold under Sandip's magnetic pull and her infatuation for the outside has rendered Nikhilesh's home derelict and lifeless is a different issue. It is, in some ways, a triumph of patriarchy in the way Tagore highlights the pitfalls of Bimala's unsolicited empowerment. Through her, Tagore toys with the possibility that the essential female nature is one of potent and destructive sexuality.

The highly troubling question that the novel seems to confront is not *how* Bimala can be liberated but whether she can be liberated without dismantling the fundamental structure of society. As a full-fledged political novel, a narrative that was generated in the crisis of this time, Tagore's stance on the role of Bimala is closely connected with the idea of nationalism. From ancient times, women have always been represented as a territory to be conquered and dominated. Equating the nation with the mother (as motherland) is an old trope in cultural practices across the world. This equation has been a source of contention among feminists because as they see it, it enables the man to control the woman and the country even as this control masquerades as protection. In India, anti-colonial movements appropriated the woman as the boundary figure to be protected against the colonial male. If the nation is gendered, then nationalism is patriarchal, as Partha Chatterjee has influentially argued.[10] In *Ghare Baire*, emergent notions of female emancipation are inextricably woven into conflicting ideas of nationalism, issues of gender identities and those of sanctioned and transgressive love/desire.[11] Just as the patriarchal voice reverberates through the declaration, "I will make Bimala one with my country", Bimala also clearly makes the woman's yearnings for freedom in the world outside her home synonymous with that of the nation waiting to be released from the shackles of foreign rule when she states:

> In that future I saw my country, a woman like myself, standing expectant. She had been drawn forth from her home... by the sudden call of some unknown... She has left home, forgotten domestic duties; she has nothing but an unfathomable yearning which hurries her on (93)

When we come to *Jogajog* (1929), we find that Kumu has waged a life-long battle against the stranglehold of patriarchy since her marriage. Though her battle against the loveless sexual hunger of her husband becomes successful in the initial stages, ultimately Tagore chalks out a history of her surrender, howsoever reluctant, to patriarchy and the unwritten laws of a male-dominated society. In fact, the end of the story can be taken as his sarcastic commentary on patriarchy itself. In the twilight years of his life, Tagore created the true modern woman – in fact the Indian version of the New Woman – in Sohini in the short story

---

[10] In *The Nation and Its Fragments: Colonial and Postcolonial Histories* (New Delhi: OUP, 2003) Partha Chatterjee notes that Indian nationalism equated the woman with the home: the spiritual and private space that is untainted by colonialism. Thus anti-colonial struggle was to ensure that this space remains sacrosanct. Women in the national struggle endorsed this view and thereby reinforced patriarchy and the nation. In his words,
"Women from the new middle class in nineteenth-century India thus became active agents in the nationalist project – complicit in the framing of its hegemonic strategies as much as they were resistant to them because of their subordination under the new forms of patriarchy"(148).

[11] See Tanika Sarkar's essay "Many Faces of Love: Country, Woman and God in *The Home and the World*" in P.K. Dutta Ed. *Rabindranath Tagore's* The Home and the World: *A Critical Companion.* New Delhi: Permanent Black, 2003:32.

"Laboratory" by uniting contradictory forces. Sohini is independent-minded, believes in women's empowerment and is not averse to sexual plurality by virtue of her sexual freedom and can, yet, stake her life for saving her husband's laboratory. She is convinced that she does not need to abide by the moral norms imposed by society. Thus, by uniting idealism and individualism, Sohini becomes Tagore's ideal woman.

From the discussion of the female protagonists of his fiction, it becomes clear that Tagore's representation of women is pre-eminently feminist, especially in the post-*Sabujpatra* phase of his writings. Though he cannot be labelled a feminist, he has always advanced the light in matters feminine. Ela in *Char Adhyay* (1934) is again a determinist who is averse to marriage in her early years so that she can chalk out the course of an independent life. But while exploring her individual personality, Tagore makes her highly susceptible to love, showing love's irresistibility.

II. Women in Cinematic Representations

The history of cinematic adaptations based on Tagore's novels and short stories has shown that almost all reputed film directors from Bengal have attempted to adapt a Tagore story on screen at least once if not more often. It seems like a rite of passage whatever the agenda in making the film might be. A survey of the films also shows that in most cases, the directors have chosen stories that are 'woman-centric' – where there is either a strong woman protagonist or a character that does not fit into the stereotypical role of the 'ideal' woman of the time. Though Tagore was inconsistent in the portrayal of strong female protagonists in all his works, he supplied film directors with a steady stream of convincing and challenging women for adapting to the silver screen. Thus adaptations like Purnendu Patrea's *Streer Patra* (1972), Satyajit Ray's *Teen Kanya, Charulata* (1964) and *Ghare Baire* (1984), Rituparno Ghosh's *Chokher Bali* (2003) and *Noukadubi* (2011), Suman Mukhopadhyay's *Chaturanga* (2009), Raja Sen's *Laboratory* (2010), to name a few, portray powerful and sometimes non-traditional woman characters as protagonists – characters who, in one way or the other, subvert the patriarchal stereotypes of the day.

Though the aim of this paper is not to make a one-to-one comparative analysis of the women in Tagore's texts and their screen avatars, it does attempt to explore the novel's possibilities in the domain of the subjective and the imaginary stemming from the director's vision of how he places it on the three-dimensional, audiovisual medium of film. Like in all cinematic adaptations, these directors often deviate from the original characterisation but it is interesting to note that in all cases, they focus visually on the uniqueness of the portrayed character as a non-traditional woman through certain visual tropes. Also, the films often devise visual semiotics that are distinctly different from the nuanced and subtle play of words that characterises the endings of Tagore's texts.

Let us first begin with "Samapti". Tagore's short story, "Samapti" [The Conclusion], is best known for its cinematic adaptation by Satyajit Ray in *Teen Kanya* [Three Daughters]. Ray's 1961 film strings together three short stories, "Postmaster", "Monihara" (The Lost Jewels) and "Samapti", written by Tagore in the 1890s. As the title of the film suggests, Ray chose to read these stories as Tagore's fictions of the feminine, focusing on the exploration of the complex 'psychology' of two adolescent girls and a young woman. Among the three, "Samapti" is perhaps the least disturbing because of its conventional 'romantic' closure and the establishment of the rites/rights of conjugality. The last scene of the film, in which the bedroom door of the couple bangs shut in the face of the bemused and nonplussed mother panting up the staircase, provides the comic closure for this carefully orchestrated film. According to Swati Ganguly, it is an appropriate 'conclusion' to diverse tales of female desire. Of course, as any Bengali reader familiar with Tagore's text knows, the short story ends thus:

> He [Apurbo] was about to get into the bed, when suddenly a pair of soft arms accompanied by the tinkling sound of bangles, encircled him in tight embrace; a pair of petal-soft lips felt upon his own like a bandit, leaving him no scope to express his surprise at the wet tear-stained passionate kiss. Apurbo, taken aback, realised that a venture once interrupted by giggles had finally found its fitting conclusion in tears. (176)

Unlike other Tagore stories in which strong-willed, sensitive, moral and emancipated women challenge the dominant patriarchal ideology of marriage, often posing radical and daring alternatives, nothing could be less true of "Samapti" in terms of its ideological underpinnings. It is conservative (not in a pejorative sense) in positing marriage as the ideal and desired condition of companionship, a realisation that slowly dawns on Mrinmoyee, the adolescent heroine when she becomes a woman capable of reciprocating her husband's romantic longing. What is interesting is the manner in which the story begins by depicting the so-called abnormality of this tomboyish village girl who celebrates joie de vivre in her manner and behaviour, flouting the norms of gendered conduct.

Coming to *Charulata*, we find that Tagore's Charu was a woman 'born' much ahead of her time. Satyajit's Ray's Charu reasserts this literary 'reality' ages after she was created in "Nastaneer" (1901). If Tagore gave life to Charu, Ray added meaning to it. It offers the maximum choice to a Bengali married woman from an aristocratic Hindu family to explore spaces though the final choice is beyond the scope of her social conditioning. The story of the film and the original literary piece unfolds in 1879, at a time when imported ideas of European Enlightenment clashed with conservative Hindu notions of home, family and gender roles. In Charu of *Charulata*, Ray discovers the crystallisation of the Indian woman, poised between tradition and modernity. Intelligent, sensitive, graceful and serene, Charu was a traditional woman whose psyche imbibed into itself waves from the world outside. Her inner rebellion is manifested throughout the narrative and cinematic spaces of the film in different ways – verbal, emotional, physical and intellectual. The film is premised on the notion of the gaze – of looking in a manner that verges on the notion of voyeuristic. Suranjan Ganguly has called it a film of 'the woman's eye' in which Charulata is given the privilege of gazing – upon the world around her and at the object of her desire – the young man who storms into her placid household and unleashes a fury of passion in her that leaves her breathless and broken at his departure. As she is empowered by a pair of opera glasses that she trains on all whom she gazes upon, it does appear that, at one level, it is her gaze that controls the film. But since Charu's gaze remains largely unreciprocated by both men in her life, it can also be stated that if looking/gazing is an empowerment, not looking back in reciprocity may perhaps be seen as an ultimate voiding of the possibility of power. All that is gained at the end of the film is knowledge, both of the self and the domestic world, by both Charu and Bhupati – a knowledge that threatens to permanently rupture the fragile web of contentment that binds them without offering any solution to the problem of disruption.

Let us consider the oft-referred ending of the film and the Tagore-Ray relation once again. Aware perhaps that there could be no visual equivalent of Charulata's poignant and rueful utterance '*na thak*' (No, let it be), Ray chose to end the film with a freeze shot, the close-up of the extended hands of Bhupati and Charu which do not meet, signifying the irreconcilable distance between them and the term "Nastaneer" rolls onto the screen. This portrays how literature does not provide space for analytical interpretations of concrete objects that present different modes of meaning in a film with respect to a given character, his/her lifestyle, interactions with other characters and their use and positioning within a film. As Shoma A Chatterjee correctly states, Ray's *Charulata*, looked at in retrospect, is a futuristic film. Time was changing and nineteenth-century Western social philosophy and Rammohan Roy's ideas were constantly working towards liberation of women. Chatterjee further analyses:

> Charu in *Charulata* celebrates a housewife's ability to *explore* spaces within the social sanction of the family she is married into; *create* spaces – emotional, cerebral, visual and nostalgic – beyond the framework of her immediate environment; *transcend* rigid spaces of family, time and place; *appropriate* space within the life and mind of another person, and *control* space when she feels she wants to. These functions are achieved within the home, (with her) husband, with the nitty-gritty of the mainly decorative 'housework' she does and with her writing. Her inner rebellion is manifest throughout the narrative and cinematic spaces of the film in different ways – verbal, emotional, physical and intellectual. She breaks every rule in the feudal family book without placing her femininity on the block.

Satyajit Ray's adaptation of *Ghare Baire* has an interesting history as it has the longest gestation period in Ray's oeuvre. Though made as late as 1984, Ray had been nurturing the idea of filming the novel since 1946, much before *Pather Panchali* emerged. Doing away with the Hollywoodian script, Ray confessed that he had been suffering from the 'pin-pricks' of his conscience for thirty-six years and so, it can be assumed that the story of the novel had been trans-created in him long before he began the actual filming. So, in the film, when Bimala appears on the screen, she is fully there unlike in the novel, where she progressively takes shape in our minds. Bimala is, of course, Ray's best depiction of the ambivalent woman. Torn between Nikhil, her cool, rational, and politically moderate husband, and Sandip, the firebrand revolutionary, Bimala shapes her way to the outer world but she is, at the same time, a victim of circumstances that bring disaster to the family. Together with her husband, she is forced to confront the wider effects of her own deficiencies and ambivalence. Though Ray's Bimala is full of tension and complexity that leads to the tragic denouement, she does not seem to be the strong-minded independent woman that Tagore depicted her as. In fact, Ray carefully juxtaposes the 'living flame' image of Bimala in the beginning with her widowed face filling the screen at the end, when the villagers bring the dead body of Nikhilesh back to the house.

In his adaptation of *Chokher Bali,* Rituparno Ghosh attempts a negotiation with the 'woman question' that occupies a central position in discourses of nationalism. At the end of the novel, we find a penitent and reformed Binodini, sobered and educated by experience, graciously forgiving Mahendra and asking forgiveness in turn before leaving for Kashi, the haven of Hindu widows. That this ending is contrived becomes clear from Tagore's own dissatisfaction with the end of the story as he stated, "Ever since *Chokher Bali* was published, I have always regretted the ending. I ought to be censured for it." Two months before he died, he wrote again, "I need to be seriously criticised for it, I deserve this criticism. I should be punished for it." For Rituparno Ghosh, this became a good opportunity to invest Binodini with agency resulting in transforming her rather abruptly and unaccountably into a feminist whose quest for autonomy merges with her search for *desh* or the homeland. This metamorphosis of rejecting the 'home' for the 'world', as mentioned earlier, invokes the binary of *ghar/bahir*, through which nationalist discourses attempted to address and resolve the question of women's emancipation. In an interview given to *AsiaSource.org*, Ghosh justifies his reason for deviating from the novel and states:

> …today when you read the novel, you can make out that this *cannot* be the ending. A lot of people wanted Binodini to get married to Behari. I think that would have been a solution 30 years ago when people were propagating widow remarriage… But in today's time, I think a woman can live on her own completely. She does not require a male surname, or title, or an appendage of any kind to help her lead her life… In the letter she writes when she leaves, Binodini mentions her own *desh*, which is not 'country', it should not be translated or read as country; it should be read as a space, a space or domain… And that is what Binodini speaks of at the end.

For Suman Mukhopadhyay, Damini and her sexual enigma appealed to him right from his college days. According to him,

> *Chaturanga*, the film, is both hated and loved, about 50:50, I would say. …It is Tagore's genius that he unified and integrated the social dilemmas that a nation was facing through the novel. I don't agree with those who say that *Chaturanga* is a novel of ideas. It is a very living text and the work of a genius and a visionary.

It is understandable that Raja Sen, in the film adaptation of "Laboratory", found it even more convenient to cast Raveena Tandon in the lead role as her broken Bengali suits the role of the Punjabi character perfectly.

Another recent film, which is again pretty woman-centric, is Pranab Chaudhuri's *Musalmanir Galpo*. Based on Tagore's short story of the same name, which is reportedly the last story he wrote, a month and a half before he passed away, it is a period piece about a young Brahmin village girl, Kamala, who's married off to a zamindar twice her age. Dacoits attack the groom's party as it is winding its way through a jungle and Kamala's husband runs for his life, leaving his bride behind. The timely arrival of Khansaheb, a Muslim zamindar on a mission to rescue young brides from dacoits, saves Kamala. Having been sheltered by a Muslim household, Kamala is now an outcast for her own family. But not one to mope over her fate, Kamala learns sword fighting and also falls in love with Khansaheb's son Karim. Happy with her metamorphosis into a woman of courage, grit and leadership skills, Khan christens her Meherjaan and pronounces her his successor by ceremonially handing her his flag. In her first fight to save an endangered woman, Kamala/Meherjaan fights the dreaded dacoit gang, forcing it to take flight and saves her cousin Sarala, who is on her way to her husband's home. The story runs on several tracks such as – a young girl's journey from weakness to strength, from powerlessness to power, from victim to victor; or, the story that talks of a time when affluent Muslim leaders rescued women in distress, their communal identity notwithstanding, and ensured they could practice their religious faith without fear or discomfort by housing them in a separate Hindu abode. Though the film is well-intentioned, it loses its way in the plethora of multiple perspectives that the director failed to take control of.

One interesting fact is that often, we find film directors drawing our attention to the radical feminist protagonist by juxtaposing her with another woman character which is just opposite to her in manner and morals. In *Charulata*, for example, we have the domesticated Manda juxtaposed against Charu; in "Streer Patra", Baro Bou is diametrically opposite to Mejo Bou (Mrinal); in *Ghare Baire*, the beautiful yet widowed sister-in-law in a white sari is portrayed against the gaudily dressed Bimala who, apart from trying on fancy blouses and taking English and piano lessons from an English lady, moves freely across the threshold of the 'home' to the 'world' outside. In *Chokher Bali*, modernity is ascribed not to Ashalata but to the widowed Binodini, who ignites passion in the men she comes across in her life to such an extent that it almost leads to the breakdown of relationships inside the home. The binary is further complicated when we see the bare-bodied Binodini moving around with more élan than the domesticated Ashalata who tries to dress in Western blouses and jackets like Bimala. The same élan is carried forward in the way the widowed and bare-bodied Damini is depicted on the screen in *Chaturanga.* In Rituparno Ghosh's recent adaptation of *Noukadubi*, the binary is once again clearly seen in the two women protagonists of the story, namely Romesh's lover, Hemnalini, and his wife, Kamala. *Noukadubi* spells out the intrigue and uncertainty of love as it grows, over time, towards the one you love. Love towards the one you happen to meet comes afterwards. It is a scathing social indictment of the times by Tagore on (a) the institution of arranged marriages where the bride and groom sometimes do not even see each other's faces before being married; (b) the blind faith in astrological horoscopes that can go completely haywire and (c) the social dictates of the Hindu family where obedient sons, after being given a 'liberal' upbringing, are emotionally blackmailed by their fathers, leading to a violation of their own principles of honesty and commitment.

To conclude, we can say that, seen from his fiction and their film adaptations, Tagore was definitely a feminist but not of the traditional kind. All his female protagonists demand some sort of 'space' or 'domain' and this makes his stories suitable material for exemplifying the 'woman question'. His views were close to what African American writer Alice Walker terms 'womanist'.[12]

Works Cited

Chakrabarti, Santosh. *Studies in Tagore: Critical Essays.* New Delhi: Atlantic Publishers and Distributors, 2004. Print.

Chatterjee, Shoma A. "*Charulata* Revisited". *8*th *Day: The Sunday Statesman.* 9 May, 2010. 1. Print.

Dasgupta, Sanjukta. "Breaking Free: Is Rabindranath's "Streer Patra" A Feminist Text?" Ed. Mohit K. Ray, *Studies on Rabindranath Tagore.Vol.II.* Delhi: Atlantic Publishers and Distributors, 2004. 181-90. Print.

Ganguly, Suranjan. *Satyajit Ray: In Search of the Modern.* New Delhi: Indialog, 2001. Print.

Ganguly, Swati and Paramita Chakravarti. "Postcolonial Negotiations with the Nation: Rituparno Ghosh's *Chokher Bali". The Indian ImagiNation: Colonial and Postcolonial Literature and Culture.* Ed. Somdatta Mandal. New Delhi: Creative Books, 2007. 242-259. Print.

Ganguly, Swati. "Gender, Sexuality and Conjugality in *Samapti.*" *Muse-India – e journal* Issue 33 (Sept-Oct 2010) ISSN 0975 1815. Print.

Ghosh, Rituparno. http://www.asiasource.org/news/special_reports/ghosh.cfm. Web.

---

[12] For detailed discussion of the film adaptation see Swati Ganguly & Paromita Charavarti's article.

—. “Riya is misunderstood.” Interview given to Priyanka Dasgupta. *Times of India.* TNN, Nov 17, 2010. Print.

Mandal, Somdatta. “*The Home and the World* as Fiction and Film.” Rabindranath Tagore’s *The Home and the World: Critical Perspectives.* Ed. Rama Kundu. New Delhi: Asia Book Club, 2001. 197-205. Print.

Mukhopadhyay, Suman. “Tagore on Screen: Filming ‘Chaturanga’” www.siliconeer.com. Web.

Nayar, Pramod. “Feminisms.” *Contemporary Literary and Cultural Theory: From Structuralism to Ecocriticism.* Delhi: Pearson, 2010. 82-120. Print.

Ray, Niharranjan. *Rabindra Sahityer Bhumika.* Kolkata, 1369 B.S. Print.

CHAPTER EIGHT

PRINTED RAINBOW: FANTASY TRAVELS BEYOND "A ROOM OF ONE'S OWN"

SWATI GANGULY

> A bespectacled, perfectly round, moon-face peeps from the dark recesses of what appears to be a medieval *haveli* or mansion. It is difficult to discern either the age or the sex/gender of this figure. What is apparent, however, is that s/he is a stranger and all agog with curiosity, an impression created through the eyes that seem to pop out. Now s/he walks along the half-lit corridors of the mysterious mansion evidently following another receding figure – a svelte damsel clad in the traditional Indian long skirt or *ghagra*, a blouse and a diaphanous veil. Her long braid sways as she sashyes seductively into a room in the splendid inner quarters of the mansion. The audience now looks through the eyes of the voyeur figure at the scene of an evening soiree that unfolds: The figures that relax on the rich carpets lean against bolsters, smoke hookah and chew *paan* are all women who belong to yesteryears – figures from a past whose tales have not been written in the annals of the middles ages. They are engrossed in the performance by the courtesan who dances to the lilting tune of a melancholy *raga;* outside, in the courtyard, naked women frolic in the water, caressing one another. It is at this juncture that the viewer-voyeur turns around and we, the audience, realise that the figure through whose eyes we have witnessed these forbidden female intimacies is an old woman.

This scene described above is from Gitanjali Rao's *Printed Rainbow* (2006); it remains vividly etched in my memory perhaps because this snippet was my introduction to this brilliant animated film which seemed to crystallise the concept of how spaces may be engendered and re-gendered. While there is nothing unusual that the secluded *andarmahal* of a medieval Indian *haveli* be inhabited only by women, there was something unique in the representation of that space as one which these women had evidently claimed as their own – a space in which they indulge in pleasures that are traditionally regarded as taboo.

Gitanjali Rao, who graduated with honours as a Bachelor of Arts from Sir JJ Institute of Applied Art, Mumbai in 1994, is a self-taught animator and filmmaker. *Printed Rainbow* premiered at the prestigious Cannes film festival in 2006 winning three awards and later went on to win several international awards. *Printed Rainbow* is a work of art which playfully toys with the ideas of re-gendering and engendering or creating spaces.

This short animation film centres around an old woman who lives with her cat in a flat in a city condominium. The camera frames her quotidian existence – a series of shots of cooking-cleaning-washing in her minimalistic household. The old woman's connection to the world outside is always from a distance: She watches those fragments of the lives of her neighbours in the condominium visible through the frame of her window. There is little to indicate that there could be anything quirky in the life of this old woman till such time as she opens her chest of drawers to reveal a collection of matchboxes. It is the 'childish' pictures on these match-boxes which triggers her daydreams or reveries. In her fantasy, she moves away from the little flat in the city condominium to strange and exotic spaces.

It was perhaps the oscillation of the film's narrative between the confined space of the flat and the open spaces of the old lady's fantasy travels which suggested to my mind the twining of two spatial tropes which are now recognised as central to feminist theory, that of 'a room' and the 'mother's garden.' As readers have already understood, my references are to the two texts which have now acquired the status of feminist literary classics – Virginia Woolf's *A Room of One's Own* (1929) and Alice Walker's essay *In Search of Our Mothers' Gardens* (1984). It is my belief that Gitanjali Rao's *Printed Rainbow* could easily be recognised as a contemporary feminist classic which has added an entirely new dimension to the relationship between female creativity and the concept of engendering and re-gendering spaces.

I

Does creativity, literary creativity in particular, have any specific relation to space? Can this relationship be understood as gendered? Virginia Woolf seemed to think so. In *A Room of One's Own* (1929), now recognised as one of the most subtle and complex feminist hermeneutics of literary creativity, she suggests that space, 'a room', is central to the identity of the woman writer. In the opening sections of this long essay, originally delivered as a series of lectures, the narrator offers an 'opinion' to her audience of expectant young women students: 'A woman must have money and a room of her own if she is to write fiction.' (Woolf 6). This money – a sum of 500 pounds a year, as we are told shortly – is a gift from the narrator's aunt. Thus in Woolf's initial proposition, inheritance and private space have to be wrested from patriarchy and owned by women to be imbued with a specific 'feminist' meaning. The idea of 'a room' evokes both a real physical space and functions as a trope for intellectual privacy and leisure. Coupled with money, it becomes a marker of the autonomy of the creative woman.

The very first chapter is a brilliant instance of how spaces can be engendered and gendered. These spaces both evoke and dismantle the neat binaries of fact/fiction, public/private, masculine/feminine, etc. The first of these is England's most prestigious academic institution which, in a sly amalgam, is fictionalised as Oxbridge: The narrator, absorbed in meditation about women and fiction, sits by the river where the willows weep and where the bushes seem to be set aflame (Woolf 7-8). Yet, this narrative, which evokes the beauty of England in late autumn with the vividness of a painting by the great 19th century British Romantic artist, John Constable, is disrupted. Poetics is replaced by the [gendered] politics of space.

As she walks across the grass, in a state of excited agitation, the narrator is interrupted by a beadle who admonishes her to move to the path since only Fellows and Scholars are permitted to walk on the grass. In the second anecdote, the narrator attempts to gain entry into the library and her way is barred by 'a deprecating, silvery, kindly gentleman'[1]. The encounter with the Beadle and Librarian drives home the point that the prestigious academic institution is a space necessarily defined by exclusion, in this case that of women. The woman writer, already famous to have received an invitation to give a lecture in the college, is, nonetheless, an outsider. The hallowed precincts of Oxbridge are an emblem of the privilege of academic status traditionally granted to men, its Fellows and Scholars.

As the narrator moves to Fernham (the fictional alias for Newnham Women's College), she is struck by the contrast with Oxbridge. Spaces are not merely about architecture and patterns of protected turf and paths but about amenities and privileges: Instead of the luncheon of sole and partridge in rich cream and wine, dinner at Fernham comprises soup, roast beef, vegetables, custard, dry biscuits and water. The leisured reverie is transformed into a sharp observation of material deprivations that mark the life of women in academia. The talk at the dinner table, as the narrator wryly observes, has flagged since 'a good dinner is of great importance to good talk.'(Woolf 20).

[1] Woolf, Virginia. *A Room of One's Own*, London: Penguin Books, 1928. Print.

It is this immediate 'material' circumstance that sparks off an animated materialist inquiry into the relation between women and public space as the narrator retires with her friend, Mary Seton, a faculty member of Fernham College, to tipple. The two women lament how their mothers and grandmothers had not left endowments for women's colleges. In their analysis, women of yesteryears had failed to do so because they did not have access to public spaces of commerce and business and hence, did not have any income. Even if they had managed, hypothetically speaking, to have 'amassed great wealth' the 'law denied them the right to possess what money they earned.'(Woolf 24).

It is not only that women have been disempowered historically, legally and politically but this very disadvantage has been 'naturalised' through the ideology of motherhood, of reproduction, care and nurture: 'Only if Mrs Seton and her like had gone into business at the age of fifteen, there would have been – that was the snag in the argument – no Mary.' (Woolf 23). With biology as destiny, many a talented woman had embraced her role as wife and mother (with an annual compulsory fecundity) and remained confined to the 'private' domestic spaces. Yet, their 'naturalised' relation to 'private' space had never been without problems for they could hardly lay claim to the home and hearth in which they cared for and nurtured their families. The gendering of the binary of the 'public/private' space as 'masculine/feminine' was thus doubly oppressive for the woman living as the 'angel in the house': Even in the 19th century, the Victorian woman never possessed 'a room of one's own.'

It is this lack that is crystallised in Chapter 4 in Woolf's survey of the real conditions in which women writers of the early 19th century wrote their fiction. She quotes from the memoir of Jane Austen's nephew who recollected how his aunt used to write:

> … she had no separate study to repair to, and most of the work must have been done in the general sitting room, subject to all kinds of casual interruptions. She was careful that her occupation should not be suspected by servants or visitors or any person beyond her own familial party. (Woolf 67)

If Jane Austen was constrained by the lack of a 'room of one's own', did a hundred years change the situation for the woman writer? In Chapter 5, there is a brief, fleeting reference to the 'bed sitting room' of an 'unknown girl' – the fictional 20th century woman novelist Mary Carmichael, whose avant-garde, but awkward fiction, *Life's Adventure*, explores the daring taboo of love and desire between two women, Chloe and Olivia. Mary Carmichael, who, as readers familiar with the text will easily recall, is the contemporary counterpart of Shakespeare's fictional sister Judith who occupies centre stage in Chapter 3: It is her tragic tale of seduction, shame and suicide, as imagined by Woolf, that had probably sparked off the debate, 'Did Women have a Renaissance?'

Yet, even at the end of this gynocritical survey, which takes in its ambit the material and intellectual life of the woman writer from the 16th century to the present, what fails to emerge is the image of the woman writer at her desk, which the idea of the 'room of one's own' would seem to suggest. As Supriya Chaudhuri has argued in her unpublished essay, *The Politics of Space: Writing in Woolf's Room*, it is 'the indefinite article' in the title of Woolf's essay to which we need to focus our attention; it is this that makes the concept of room, 'indefinite, that is to say, hypothetical, unrealised... it is a space that, paradoxically, never takes shape in a text that is almost entirely about space and place. The text gestures towards it, returning over and over to the room where Woolf herself sits at her desk to write but the two never quite merge.'

How does one interpret such a textual strategy of deferral and denial? Does Woolf deliberately tease her audience/reader with a promise to revoke it completely? Or is one missing out on something crucial? Towards the end of Chapter 1, nestling quietly, inconspicuously amidst the flow of recollections, is a startling sentence – 'and I thought how unpleasant it is to be locked out; and I thought how it is worse perhaps to be locked in.' (Woolf 25).

If women have traditionally been excluded – 'locked out' – from the privileged sites in patriarchy, the very same system has also endorsed and sanctioned that any unruly woman, a woman who transgresses codes of femininity, can be punished by being 'locked in'. To be 'locked in a room', thus, carries the connotations and threats of violence often reserved for women, the transgressive or subversive daughter or wife. It evokes, to the reader's mind, Charlotte Bronte's *Jane Eyre*, that classic Victorian fiction of bourgeois female 'emancipation' which is predicated, ironically, on the real incarceration of Bertha Mason – the wife of the masterful Rochester; a madwoman, reduced to the condition in Thornfield Hall. The 'redroom' in Gateshead where the young Jane is locked in for being uppity, evokes from the same text, the attic of Thornfield Hall where the 'mad' Bertha Mason/ Rochester is incarcerated. Rooms with locks carry a dangerous potential – of perilous slides from a space free of interruptions and of leisure to the space of prisons/madhouses. Though these spaces are never mentioned in Woolf's essay, they hover on the horizon of her narrative and train of thought, in her indictment of Charlotte Bronte, who wrote foolishly and in rage, a woman who 'was at war with her lot.'[2] It is evident that the notion of a secluded 'feminine' space, 'a room' could hardly be a felicitous one for the avant-garde Bloomsbury writer who believed that the artistic mind must be 'androgynous'.

So, at the end of her essay, Woolf does not reject the notion of 'a room of one's own' and money but recasts them as conditions for the collective future of woman writers: 'For my belief is that if we live another century or so… and have five hundred a year each of us and rooms of our own… then the opportunity will come and the dead poet who was Shakespeare's sister will put on the body which she has often laid down.'(Woolf 112). In Woolf's polemics, the space of 'a room' – a hypothesis and a promise – can emerge only if there is a collective will and effort to 're-gender' it, to free from its associations of a private/feminised/secluded space with ambiguous potentials to a space which will serve as the vantage from which to engage with 'the common life'. It is this which will engender the promised 'room' which will foster the 'habit of freedom and the courage to write exactly what we think.' (Woolf 112).

II

So powerful is Woolf's essay in creating a spatial template of female creativity that feminists have continued to engage with it, if only to challenge its very premise. The obvious instance of this, as we all know, is Alice Walker's 'womanist' essay *In Search of Our Mothers' Gardens*. If Woolf's *A Room of One's Own* has provided *the* spatial trope for female creativity in feminist literary theory, Walker's essay challenged the white European middle class bias underlying Woolf's feminist thesis from the perspective of the black woman artiste. Thus Walker posited 'mothers'gardens' as the alternative space in which black women artistes nurture their creative genius. This shift is necessarily also a redefinition of what constitutes artistic activity – from the high culture of Woolf's Bloomsbury circle to the fields where African-American slave women were forced into hard labour; from the novel as the preferred genre of female creative expression to the songs and tales as forms of oral creativity that, according to Walker, comprise the black African-American myths; instead of the single, unmarried women who serve as Woolf's models of the creative writer, Walker posits mothers, whose bodies have been broken by bearing many children, and yet who have managed to retain their spirit and deep spirituality.

---

[2] Walker, Alice. "In Search of Our Mothers' Gardens". *In Search of Our Mothers'.*

Walker critiques the confident assertion of the demand for inheritance and private possession uttered by the specially privileged position of a white, middle class liberal. Indeed, *A Room of One's Own* has its own politics of exclusion, albeit unintentional, in its failure to envisage even the possibility of the 'creativity' of the black woman. Walker describes, in moving, almost Sibylline prose, their destinies (as they are) 'sold into slavery and mutilated in body… dimmed and confused by pain' and still retain in their hearts a spirituality so deep that it manifested itself in the songs of the Blues singers.[3]

Her strategy of countering Woolf's 'colour blindness' is two-fold: In the story of Phillis Wheatley, the young slave girl who was torn by 'contrary instincts', she makes direct textual interventions, as parenthetic comments, as she quotes extensively from Woolf's text:

> Virginia Woolf wrote further, speaking of course not of our Phillis, that "any woman born with a great gift in the sixteenth century [insert "eighteenth century", insert "black woman", insert "born or made a slave"] would certainly have gone crazed, shot herself, or ended her days in some lonely cottage outside the village, half witch, half wizard [insert Saint]… For it needs little skill and psychology to be sure that a highly gifted girl who had tried to use her gift for poetry would have been so thwarted and hindered by contrary instincts [add "chains, guns, the lash, the ownership of one's body by someone else, submission to an alien religion"] that she must have lost her health and sanity to a certainty. [Walker 235]

In the second instance, she retells the story of her mother who lived exactly in the time when Woolf was lecturing the women students of Fernham. It is here that we have the alternative spatial trope of female creativity: The open spaces of the garden where her mother planted 'fifty different varieties of plants that bloom profusely from March until late November' (Walker 241). The notion of women's creativity as Walker's essay so powerfully suggests, cannot be confined to 'writing'; one has to search for it in other archives – songs, oral narratives, the quilt in the Smithsonian Institute which conveys to the visitor the presence of the anonymous black woman artist 'of powerful imagination… '

Woolf and Walker, disparate in their politics, and thus the poetics of space, seemed to happily co-exist in Gitanjali Rao's animation film *Printed Rainbow* (2006). Perhaps there was a common thread after all; the very fact that both writers end their essays with the promise of the artiste

---

[3] Walker, Alice. "In Search of Our Mothers' Gardens". *In Search of Our Mothers' Gardens.* London: The Women's Press, 1984. 232-34. Print.

within us – the ordinary woman who, like this old lady, partakes of the 'common life' and has the potential of transforming it in her imagination.

One does not even have to be a maker of cultural artefacts to be an artist as Gitanjali Rao's film so powerfully suggests. The film has two major episodes of dream-reverie. In the first one, the old lady's imagination is sparked off by the picture of an elephant on a matchbox. The very form of the animation film allows a seamless slippage from the black-and-white space of the city to the old lady's dream-reverie. In her fantasy, the old lady rides on the elephant's back through lush-green mellow landscape, makes a journey by a boat which takes her to the doorstep of a 'medieval' Indian all-female *haveli.* The voyeuristic experience of a peep show into a feminist utopia leads to a more overtly liberating one; the sight of a matchbox with the picture of a truck induces an entirely different fantasy: she drives a truck through highways, speeding to overtake other vehicles, enjoying the breeze in her hair.

It is at this moment that her day-dream is interrupted by the sound of the doorbell. She has a visitor, an elderly gentleman who has a dog for a companion. They drink tea in companionable silence after which the old gentleman, evidently another avid collector of matchboxes, gifts her one with the picture of an idyllic cottage with a 'rainbow' printed on it. This particular matchbox allows her to create her final travel in fantasy to a never-never land. The old lady and her cat move to this animated magical countryside – a cottage by the stream, a garden which is always in bloom and a picture-perfect farm. Happy and contented in her domesticity in this new liberating space, at the end of the day, the old lady reclines in her rocking chair with her cat on her lap. When the doorbell rings this time, she does not wake from her dream-reverie. Indeed, she is never to wake up from it.

In the final magic-real shot, her friendly neighbour discovers her in the picture on the matchbox titled 'Printed Rainbow' rocking contentedly in her armchair with her companion cat. She has been liberated in her death to the world of her dreams. In Rao's film, spaces are 'engendered' through an act of re-gendering and claiming them as a woman's own.

Works Cited

Rao, Gitanjali, dir., *Printed Rainbow* 2006. Film.

Walker, Alice. "In Search of Our Mothers' Gardens". *In Search of Our Mothers' Gardens.* London: The Women's Press, 1984. 232-34. Print.

Woolf, Virginia. *A Room of One's Own*, London: Penguin Books, 1928. Print.

CHAPTER NINE

SPATIALISING FAMILIES AND FANTASTIC COUPLES: PHOTOGRAPHS AS CULTURAL IDENTITY AND "PERFORMANCE"

HARDIK BRATA BISWAS

Pic 1. K. C Banerjee & Uma Banerjee. 1930. Kolkata (7.98" x 9.94").

Pic 2. K. C. Banerjee & Uma Banerjee. 1930s/1940s. Kolkata (5.49" x 3.21").

Two photographs of KC Banerjee and Uma Banerjee, sourced from a family related to the Sovabazar *rajbari* in north Kolkata, neither depict the splendour of the *zamindari* family nor a royal couple. They do not even attest to a reification of the high elitism of the late eighteenth and nineteenth centuries as found in paintings or formal studio techniques used for 'royal' sitters much later. On the contrary, the props – a crescent moon and an airplane – give the viewer a heightened sense of the 'modern' both in terms of the progress of conjugality and the advancement in studio techniques from the late nineteenth to the early twentieth century. Taken in the 1930s and the 1940s in Kolkata studios, the two photographs of the same couple show a comparatively more liberal choice of posture and backdrop than earlier formal presentations of newlyweds.

This is not to make an exceptional case for the two photographs selected but to treat and read them as symptomatic of the changing visual worlds of identity and representation vis-à-vis conjugality and family in the early decades of the twentieth century and see how questions of performance and identity are linked with the histories and cultures of power that led to their archiving in the family album. Before approaching particular photographs, let me go back to the structures that produced photographic images of women of the region.

The invention of photography and the arrival of nineteenth-century modernity were incidents that not only coincided with historical and ideological developments of the era but were also steeped in debates that either defended emerging technologies or attacked newfound modes of representation. Modernity, taken as a rupture with traditions, acquires a position of continuity that ushers in new modes of representing knowledge and experience.

With the advent of the modern age in Europe, systems of visual representation underwent radical changes in technical and aesthetic spheres, particularly in the field of photography. The influence of modernity in various spheres of life conditionally redesigned existing norms of representation with the visual form appearing as a primary category where shifts began to happen. The development of photography, its increasing popularity, the emergence of compact families and changing roles of women are interlinked in a mesh of macro and micro changes in the realm of the social and the technical.

The desire to have a photograph of someone who was away from home was historically conditioned by a redefining of the concept of family, brought about primarily by the Industrial Revolution in Europe. It was intensified further by the rise of modern capitalism that destabilised the structure of the feudal self-contained landed gentry. For the first time in the history of mankind, 'going away' to sell one's labour became common. Visual representations of women began to emerge in the light of the newly-constituted, smaller families and gender-based division of labour was justified by the dominant male agency through generalised divisions between home/world, private/public and female/male.

Crary shows that the invention and rising popularity of photography as a medium and other related forms of 'realism' in the nineteenth century have been part of a continuous unfolding of a Renaissance-based mode of vision in which photography, and eventually cinema, are part of a continuing perception of space in perspective.

Crary argues that it is not enough to attempt to describe a dialectical relation between the innovations of avant-garde artists like Manet and writers in the late nineteenth century and the concurrent 'realism' and positivism of scientific and popular culture. It is crucial to see these phenomena as overlapping components of a single social movement which modernisation of vision had begun to impact decades earlier. He suggests that a broader and far more important transformation in vision occurred in the early nineteenth century. Modernist paintings in the 1870s and 1880s and the development of photography after 1839 can be seen as later symptoms or consequences of this crucial systemic shift, which was well under way by 1820.[1]

Changes in the intelligibility of the visual medium were inseparable from a 'massive re- organisation of knowledge and social practices that modified, in myriad ways, the productive, cognitive, and desiring capacities of the human subject'.[2] The early work of Jean Baudrillard details some of the conditions of this new terrain in which a nineteenth century observer was situated. For Baudrillard, one of the crucial consequences of bourgeois political revolutions at the end of the 1700s was the ideological force that animated myths of the rights of the man and the right to equality and happiness. In the nineteenth century, for the first time, observable proof became necessary in order to demonstrate that happiness and equality had, in fact, been 'tainted' after watershed struggles that gave birth to an essentially non-elite culture. This was, to a large extent, possible, as Benjamin pointed out in his much-discussed thesis on

---

[1] Jonathan Crary, *Techniques of the Observer: On Vision and Modernity in the Nineteenth Century* (Massachusetts: October Books, 1992) p. 5.

[2] Crary, 3.

the condition of the work of art in the age of mechanical reproduction[3] because 'happiness' had to be measurable in terms of objects and signs.[4]

Developing on later twentieth-century advancements in the sphere of mechanical reproduction techniques, which would eventually gather immense potential in reshaping the economy of visual representation and other forms of representation, Walter Benjamin argues in favour of commodity generating a 'phantasmagoria of equality'. Unfortunately, in spite of promises of 'equality', the medium was destined to emerge as a phantasmagoria of representation within the capitalist format of economy. With the emergence of cheaper and easier means of producing exact reproductions, imitations, copies and counterfeits, techniques of producing them posed as challenges to aristocratic monopoly and control over signs which were part of traditional patriarchy.

The technique of mimesis generated social power based on the capacity to produce equivalences. Baudrillard further writes that it is clearly in the nineteenth century, beside developments of new industrial techniques and new forms of political power, that a new kind of sign emerges.

These new signs, 'potentially identical objects produced in indefinite series', herald the moment when the problem of mimesis disappears – "The relation is neither analogy nor reflection, but equivalence and indifference. In a series, objects become undefined simulacra of each other…"[5] What Walter Benjamin called the 'cult' value of the picture was effectively abolished when photographs became too common to be remarkable; they were objects to be consumed and thrown away.

---

[3] Walter Benjamin, *The Work of Art in the Age of Technological Reproduction (Second Version)*. In Michael W. Jennings, Brigid Doherty, and Thomas Y. Levin edited The Work of Art in the Age of Its Technological Reproducibility, and Other Writings on Media (Massachusetts: The Belknap Press of Harvard University Press, 2008). This version was written between late December 1935 and beginning of February 1936 while in exile in Paris. Unpublished in this form in Benjamin's lifetime. As observed by Michael W. Jennings in the short introduction to the section I of the volume, Benjamin defined the field on which his theses on modern media will move. Benjamin mainly addresses two questions: 'the capacity of the artwork to encode information about its historical period…and the way in which modern media – as genres and as individual works – affect the changing human sensory apparatus.' (Jennings et al eds. 2008: 9)

[4] Jean Baudrillard, Jean Baudrillard, La société de consommation, 60. Gallimard, Paris 1972 – English translation in Jean Baudrillard, Selected Writings (Ed. Mark Poster) (Stanford University Press, Stanford 1988) pp. 29-56.

[5] Jean Baudrillard, *L'echange symbolique et la mort* (Paris, 1976), 86.

According to Tagg, a more far-reaching pictorial revolution had taken place: The political axis of representation had been entirely reversed.[6]

Being photographed was no longer a privilege. Beyond the apparent democracy of the medium, it was soon to become a new kind of power at the disposal of the social body, generating new kinds of knowledge and newly refined means of control. Crary uses Baudrillard's argument to show that, amongst these new fields of serially produced objects, the most significant, in terms of their social and cultural impact, were photographs. As Crary says,

> The photograph becomes a central element not only as a new commodity economy but also in the reshaping of an entire territory on which signs and images, each effectively severed from a referent, circulate and proliferate. Photographs may have some apparent similarities with older types of images, such as perspective paintings or drawings made with the help of camera obscura; but the vast systemic rupture, of which photography is a part, renders such similarities insignificant. Photography is an element of a new and homogenous terrain of consumption and circulation in which an observer becomes lodged. To understand the 'photography effect' in the nineteenth century, one must see it as a crucial component of a new cultural economy of value and exchange, not as a part of a continuous history of visual representation.[7]

Photography in its 'modern' form emerged through technical, ideological and aesthetic reconfigurations right from the early nineteenth century in Europe. It was shortly, almost simultaneously, to arrive on Indian shores.

It was also significant that the camera intervened in the private sphere of the family in the late decades of the nineteenth century and went on to a represent what was public within the private. The reciprocity between public and intimate spheres and an emergence of public spaces within the private domain is a major area of colonial photography. In the context of colonialism, photography, as one of the technologies facilitating memory, supplemented nostalgic remembrances of the individual by the reproduction of her/his own past. The colonised body and a desire to possess a new individuality and an unconscious obsession with tradition signify the new social formation of colonial modernity.

Colonial modernity, in which the popularity of the family album was witnessed, was not an antithesis to tradition. It was a mutual coexistence

---

[6] Tagg, John. *The Burden of Representation: Essays on Photographies and Histories.* (London: Macmillan, 1988) pp. 47-59.

[7] Crary, 13.

with tradition. It can, at best, be a twofold representation of bourgeois ideology and tradition.[8]

Science, as an agency, 'traversed a vast arena, encompassing fields from literature to religion, economy to philosophy, and categories from elite to popular. Such divisions overlook and conceal how politics and religion, and science and the state, run into each other and how, through spillovers and transgressions, modernity penetrates the fabric of social life'.[9]

Thus, women have been subjected to representations as part of a predominantly male gaze and 'private' modes of domesticity, deified repression of women and peace in the family could now be copied and preserved as documents of indigenous progress with the new technology simultaneously keeping intact the 'honour' of women.

Gendering and power could actually be played out in front of the camera with regard to positioning, dressing and choice of co-sitters for women. The conquered space of domesticity could now be put under closer inspection of the institutions of conjugality and notions of romantic love which facilitated rather than coercively restrict the Bengali middle class.[10]

From the 1870s to the early twentieth century, photography played a significant role in constructing the social and family history of the region. To Malavika Karlekar, adding a visual element to the history, context, or even relationship, that may have been written about, discussed, and analysed, is not merely to slip in some 'pretty or telling' images. Rather, she argues, 'reading' photographs is to introduce another dimension to the experience of colonialism in Bengal.[11] With the advent of colonial modernity and its institutions of education, professionalism and commerce, there was a 'controlled blurring of the limits of racial divide.'[12]

---

[8] Sujith Kumar Parayil. "Photography and Colonial Modernity in Keralam" in *Space, Sexuality and Postcolonial Studies* edited by Manas Ray. (Kolkata: Centre for Studies in Social Sciences, Calcutta, 2003) pp. 97-120.

[9] Gyan Prakash. *Another Reason: Science and the Imagination of Modern India* (New Delhi: Oxford University Press, 2000) p.7.

[10] For a detailed account see Tapan Raychaudhuri.2000. Love in a Colonial Climate: Marriage, Sex and Romance in Nineteenth-Century Bengal. *Modern Asian Studies* 34, 2 (May) 349-78 and Sambuddha Chakraborti.2003. Conjugal Relations in Early Nineteenth-Century Bengal: Conditions before the Growth of a Private Relationship in Anuradha Chnada, Mahua Sarkar et al. edited *Women in History*. Kolkata: Progressive Publishers. 95-129.

[11] Malavika Karlekar, *Re-visioning the Past: Early Photography in Bengal 1875-1915* (New Delhi: Oxford, 2005), 4.

[12] Karlekar, 4.

From the 1870s to the early decades of the twentieth century, photography played a significant role in the construction of the social and family history of the region. The *bhadro samaj* accepted photography as a useful innovation that fractured the public/private dichotomy without too much dislocation. The varied degree of exposure of public in private and private in public through photography also facilitated, to some extent, the task of social reformers, the educated Bengali middle class and the *baboos*, to make expose the women in their families. For the *bhadrolok*, photography was useful in depicting family life and the professional roles that they had newly acquired.

Photographs recorded individuals and their prosperity; they also portrayed the emergence of the conjugal unit, especially the woman, who was no longer expected to observe the rigours of *pardah*.[13] Structures like the family, girls' education and women's emergent roles were, to some extent, helped by the dynamics of the new visual space of the studio and the camera. The debut of photography into Bengali middle class homes from the 1870s can be seen as an affiliation between the colonised and coloniser, where the technology of photography travelled into Bengali homes in a spirit of assimilation.

From the mid-90s, sections of the *bhadro samaj,* which were increasingly moving away from land-based occupations to professions like government service, law, medicine and teaching, were simultaneously redefining their lifestyles. Contrasting the modern rational outlook with these changes was important in the process of refiguring, in which women emerged as a primary exhibit. Shifts in patterns of visual representation were one among many other shifts.

Changes in lifestyle, family arrangements, in modes of dress, demeanour and social relations, were going hand in hand but were fettered by conflicts with regard to class positions. For the evolving nationalist paradigm in the nineteenth century, it was necessary to cultivate the material techniques of Western modern civilization while retaining and strengthening what was projected as the distinctive spiritual essence of national culture. The methods for representing the new middle class family were drawn from a variety of sources: A reconstructed 'classical' tradition, modernised folk forms, the utilitarian logic of bureaucratic and industrial practices, the legal idea of equality in a liberal democratic state, were among these sources. Dichotomies within these discourses were many and cohabiting. The emergence of such a woman involved not only adjustments in traditional mindsets but also appropriate education, training

[13] Malvika Karlekar. *Visualizing Indian Women 1875□1947*. (New Delhi: Oxford University Press, 2006)

in household management skills, dress reform and increased visibility in social spaces.[14]

In 1856-57, Calcutta was the focus of a growing public sphere and location for cultural interactions beyond the family. It witnessed the emergence of the new white-collar worker. The home too, became an arena for change and adjustment. With the camera becoming a vital tool for recording roles and relationships, photographers gradually replaced portrait painters.[15]

As Karlekar observes, 'With changes in family dynamics, it became imperative for [the] woman to visit the studio as well. Public viewing through the photograph, so to speak, of mothers, wives, and sisters, was integral to the affirmation of emerging identities and an engagement with gender relations. In a formerly *pardah* society, visibility of the woman became a metaphor of her changing status. And as Calcutta increasingly afforded the opportunity for new definitions of the household, family, conjugality, parenting and childhood, the camera was available to record them.'[16] [17]

---

[14] Karlekar, 93-94.

[15] Beside studios and professional photographers, early pioneers in Bengal were Birchandra Deb Barman Manikya, the Maharaja of Tripura, Manmathanath Ghosh, Samarendrachandra Deb Barman, Nilmadhav Dey, the founder of The Bengal Photographers (1862), Charuchandra Dey. The Calcutta Art Studio established by Annadaprasad Bagchi (1878), Motilal Nag (father of Kalidas Nag), Gaganendranath Thakur, Upendrakishore Ray, Kuladaranjan Ray of the same Ray family, Hemendromohan Basu, more popularly known as H. Bose, Abinash Dhar, Parimal Goswami, Ramananda Chattopadhyay of *'Prabasi'* and *'Modern Review'* fame, D. Ratan and Company, Charuchandra Guha (C. Guha studio at Albert Hall, College street, Kolkata) and others were pioneers of photography in this region.

[16] Karlekar, 94.

[17] How does vision structure the day-to-day practices of family life? How does it produce family relations and form family memory? An inquiry into familial looking must begin by exploring the ways in which the family is inscribed within a system of representation that inflects its construction. It needs to examine the elements of such representational systems, the particular historical circumstances and conventions that shape them. Close readings of commonplace visual images and how women engage with them vis-à-vis memory such as family photos, as well as particular scenes of looking and particular photo sessions within a family's history can illuminate the working of familial gaze and familial looks. Such readings can also reveal the particular social and cultural screens that mediate these looks, infecting the familial gaze with class and sexual difference.

Pic 3. P.C. De and Snehalata. 1903.

Pic 4. Promod Chandra Palit and Maya Palit. 1923, Kolkata (7.37" x 10.34").

In most formal or informal couple photographs from the late nineteenth and early twentieth century, the couple was seen blissfully separated from the complex and restrictive formation of the joint family system. The newlywed couples made best use of photography and the studio space 'to create public and formal representations of the bourgeois couple from among themselves'.[18] The romantic freedom in everyday lives and the ideals of marriage, that were not so easily found in the space of the reconstituted joint family at the turn of the century, established a place of representation in the widely popular photographic images of the conjugal couple. On one hand, the ideal of the nuclear family where the husband and wife share a companionate space, and on the other, the everyday reality of the household with other members of the family where interactions between the husband and wife were few and far between, made couple photographs especially popular to the Bengali middle class.

Popular displays of such photographs inside households, bedchambers and in lockets speak of the growing popularity of this form in the early decades of the twentieth century. Formal couple photographs were probably the first of their kind to venture out of the pages of the family album and find other places of independent display.

With the growing popularity of such marriage memorabilia over the years, expressions gradually became relaxed, giving the viewer a sense of the 'expected' bonding between the pair. The couple photograph, while doing its bit to attest to the conjugal identity of the subject with respect to

[18] Rochona Majumdar. *Marriage and Modernity: Family Values in Colonial Bengal* (New Delhi: Oxford University Press, 2009) 127.

the other sex in the frame, grew to popularity at a time which was fraught with differences between reformist projects and the Hindu revivalism at the end of the nineteenth century. Although poets and writers at the end of the nineteenth and the beginning of the twentieth century valorised romantic love and ideals for the conjugal couple in their works, there was also the emergence of a growing body of advice manuals for women, pamphlets, poetry and farces that rebuked romantic ideals and the supposed freedom of the couple, especially of the woman, as destructive for the community and the nation alike. Hence, in a way, negotiating everyday criticism was far more difficult for the couple and especially for the woman in the otherwise happy frame of conjugality.

The history of visualisation in Bengal with the coming of English education and colonial modernity is not linear. Reception and responses to patterns of modernity and social changes were rupturous and never defined well. Home was soon at the focus of these changes. Partha Chatterjee has pointed out that in nineteenth century Bengal, the need to modify social relations and religious practices was often a burden for men, a necessary obligation. His argument is that it fell upon women to maintain the purity and sanctity of home. As it happened, the domain of the family and the position of women underwent considerable changes. It was undoubtedly a new patriarchy that came into existence, different from the 'traditional' order but also explicitly claiming to be different from the 'Western' family. The 'new woman' was to be modern, but she would also have to display the signs of national tradition and therefore would be essentially different from the 'Western' woman.[19] In this respect, the colonial state as a modern regime of power did not prohibit, but facilitated the process.

The figure of the woman often acts as a sign in discursive formations, standing for concepts or entities that have little to do with women in reality. Each signification of this kind also implies a corresponding sign in which the figure of man is made to stand for other concepts or entities, opposed to and contrasted with the first. However, signs can be operated upon, connected to, transposed with and differentiated from other signs in a semantic field where new meanings are produced.[20]

For the nationalist paradigm, it was necessary to cultivate the material techniques of modern Western civilization while retaining and strengthening what the former projected as the distinctive spiritual essence of national culture. This called for a selective appropriation of Western modernity. The solutions for tensions in which the new middle class

---

[19] Partha Chatterjee, *Nation and Its Fragments* in The Partha Chatterjee Omnibus (New Delhi: Oxford University Press, 1999), p. 9.

[20] Chatterjee, 68.

family found itself were drawn from a variety of sources: A reconstructed 'classical' tradition, modernised folk forms, the utilitarian logic of bureaucratic and industrial practices and the legal idea of equality in a liberal democratic state.

To quote Chatterjee: The new woman defined in this way was subjected to a new patriarchy. In fact, the social order connecting the home and the world in which the nationalists placed the new woman was contrasted with not only that of modern Western society; it was explicitly distinguished from the patriarchy of indigenous tradition. The new 'woman' was quite the reverse of the common woman, who was coarse, vulgar, loud, quarrelsome, devoid of superior moral sense, sexually promiscuous, subjected to brutal physical oppression by males. The making of the *bhadromahila* as a compassionate category of the *bhadrolok* was dependent, amongst many other things, on a 'feminine curriculum'.[21]

As corollaries of these somewhat unstructured social formulations and strictures, representation of women became extremely important within the parameters of ideology. Family, a well-structured home and conjugal relationships emerged to be a prime locus of photographs. A *new family* began to be structured with the compulsions of urbanisation, office jobs and limited living space. As in Britain, the new family came forth as a private domain. The companionate wife was expected to act as a buffer between the pressures of a fast-changing outside world and the home. To quote Chatterjee: The simple historical fact is that the lives of middle class women, coming from that demographic section that effectively constituted the 'nation' in late colonial India, changed most rapidly precisely during the period of the nationalist movement indeed, so rapidly, that women from each generation in the last hundred years could say quite truthfully that their lives were strikingly different from those led by the preceding generation. These changes took place in the colonial period mostly outside the arena of political agitation, in a domain where the nation thought of itself as already free."[22]

Even if the nationalist discourses in Bengal tried to devise a monolithic identity for women, identity remained ruptured. Performances of the self before the camera can make representation go beyond the grasp of the normative structural economy of framing. As representational spaces, studios and photographs opened up possibilities of identity formation and performance of the subject in a changed historical context. Economic, social and political events in the region and related forms of knowledge, in turn, produced new sensibilities for people within parameters of social

[21] Chatterjee, 127-128.
[22] ibid, 132-133.

spaces. It is where the inter-subjective communication and negotiation is made possible for the subjective refashioning and invention of the modern self.

Rational or imaginative, emotional histories also shape perceptions of people through its workings in social spaces.[23] The process of consumption, appreciation and retrieval of memories through photographs is often incompatible with dominant modes of representation in the public domain. Unexpected ruptures and overlooked possibilities appear in the play of memories and desires. The aesthetic of displaying, arrangement of photographs cannot always be subjugated by preponderant dictums of popular aesthetic choice in a specific time within colonial/post-colonial urban space. The subjective nature of photographic albums, each with its distinctive narrative, does not usher in a closure of the possibilities of the visual but also open up the scope for reading patterns of gendering. The non-public arena of family albums gives it a chance to preserve moments in it which, otherwise, can be prejudiced in spaces beyond the limits of kinship ties within which such images of women are subject to discursive exchanges.

Viewership of family albums is usually restricted in the limited and porous domain of familial kinship. History of families, uprooting, and reconstitution are relived through the photographic experience. Stuart Hall suggests that cultural identity is not something that exists in a performed state, transcending time, peace and history. Rather, it is always in a process of constant transformation.[24] Identity is not something fixed in the past, awaiting discovery; nor is it a crystallised essence waiting to be uncovered. On the contrary, it is subject to the movements of history, culture and power. However, Hall also points out that cultural identity is not merely a question of whim and imagination, since all identities have their histories, and histories have real effects, both material and symbolic. In addition, they are always constructed through memory, fantasy, narrative and myth. Cultural identity is an unstable point of identification constructed through speech, acts and artifacts. An identity is not an essence, but a positioning. Probing into lives through photos can be

[23] According to Lefebvre (1991, 33, 73) social space can mean a subjective space where the subject's actions are intertwined with negotiations as well as claims for methodical reforms. It is also the outcome of past action and at the same time it permits fresh actions to occur, while suggesting others and prohibiting some. The nature of the social space as a dense fabric of network and channels attached to the everyday has also been observed by Lefebvre (2002, 231).

[24] Stuart Hall, *"Cultural Identity and Diaspora"*, in Identity, Community, Culture, Difference, ed. Jonathan Rudeford (London: Lawrence and Wishart, 1990).

difficult. Despite their documentary power, their status as historical evidence has been the subject of intense debate.

The notion that a photo can summon the past in any unmediated, transparent way is now generally regarded as a naïve concept. Albums can serve as both instigators and preservers of memory. They embody a desire to preserve the external image, which helps remembrance. For Walter Benjamin, memory functions in such a way that images simply accumulate in no apparent order. There is no chronology that orders this succession of memory flashes. Benjamin argues that an image acquires meaning to the extent that it finds a corresponding moment in the present.[25] Here, it would not be entirely out of place to mention that in some of the most incisive writings about emulsion-based photography, the image is seen as a shock to the flow of time. Photographic camera, in Walter Benjamin's opinion, interrupts the ordinary continuum of history. Photographs bring death to the photographed, but precisely in transforming history into a cemetery and in converting the past into a spectre haunting the future, a photographic image can also stimulate serious solidarity between the dead and the living.

Photographic images, according to Benjamin, turn time into space. They disclose that which has been forgotten or overlooked in historical time and in this way, they no longer allow us to think of history as continuous and linear or to see our present as a mere reproduction of the past. Arresting temporality, photographic images establish and verify material connections across time: They display history as a discontinuous site of 'magic', 'epiphany', 'correspondence' and 'shudder'.[26]

To quote Susan Sontag, "Each photograph is only a fragment, its moral and emotional weight depends on where it is inserted. A photograph changes according to the context in which it is seen".[27] Sontag's remark calls attention to the fact that emulsion-based photography is not as closed, homogeneous and cut-and-dried as many of its commentators (including, at times, Benjamin), often projected it to be. Meaning and reference, according to Sontag, do not reside in the photograph itself. They emerge

---

[25] Walter Benjamin, *"Peguena historia de la fotografia" in Discursos ininterumpidoes,* vol. 1, Filosofia del arte y de la historia (Buenos Aires: Taurus, 1971), quoted in Daniel James and Mirta Zaida Lobato, *Family Photos, Oral Narratives and Identity Formation: The Ukranians of Berisso* in Hispanic American Historical Review 84.1 (2004), 5-36, p.13.

[26] Lutz Keopnick, *"Photographs and Memories"* in South Central Review 21.1 (2004) pp. 94-129.

[27] Susan Sontag, *On Photography* (New York: Farrar, Straus and Gironx, 1977), pp. 105-6.

from what we do with certain images, from how we invest them with knowledge and desire, from how we allow them to affect the viewer under particular circumstances and position them against the background of competing discourses.

The complex question of performance surrounds couple photography, simultaneously raising questions about identity formation and performance. The choice of the backdrop cannot be fully understood only from the point of view of the material choice the couple had in the studio they visited. The space where the props and the backdrops transfer the couple is also a real space where they aspire to soar above the mundane in an airplane or choose the crescent moon as the seat for their romance. A conscious self-presentation of an ideal nature in the performance space of the studio not only represents internal aspirations but also an aspiration to be recognised as distinctive social subjects.[28] The ideal values such as expression of wealth, luxury and class are incorporated in their performance as observed by Goffman.[29]

The objectivity of the photographic process lends it a mass credibility within structures of visual modernity. To quote Bazin:

> Only a photographic lens can give us the kind of image of the object that is capable of satisfying the deep need man has to substitute for it something more than a mere approximation, a kind of decal or transfer. The photographic image is the object itself, the object freed from the conditions of time and space that govern it. No matter how fuzzy, distorted or discoloured, no matter how lacking in documentary value the image may be, it shares, by virtue of the very process of its becoming, the being of the model of which it is reproduction; it *is* the model.[30]

The two photographs of the Gangulys on one hand borrows from the western neighborhood studio tradition with regards to the choice of the settings they chose as evident from the comparison with the following early twentieth century photographs from the United States.

The two photographs of the Gangulys on one hand borrows from the western neighborhood studio tradition with regards to the choice of the settings they chose as evident from the comparison with the following early twentieth century photographs from the United States.

---

[28] Parayil, 284.

[29] Erving Goffman. [1959]. The Presentation of the Self in Everyday Life. (London: Penguin Books, 1990) 28-50.

[30] André Bazin, *What is Cinema?* translated H. Gray (California: University of California Press, 1992) p.14.

Pic 1 & 5 Margaret Ballardini & Fred Watson, Luna Park, St Kilda, Victoria, CA. 1927. Photographed by M D True, Electric Studio.

Pic 2 & 6 Painted photographer's studio backdrop of airplane used for Aerial portraits.

On the other hand, representations like photographic postcards and print literature invoke the history of mentalities of the time. The rise of the scientific and the fantastic and a happy mixture of both in print literature in the early decades of the twentieth century had informed the mind of the Bengali middle class. Exploration, obviously removed from the colonial master's use of the word, had been a major trope in the finding and projection of the self in the cultural milieu of the Bengali middle class. Apart from translations of works of Jules Verne, Jonathan swift and Daniel Defoe, travelogues, diaries of expeditions and popular essays on natural sciences, the genre of scientific romance was also popular. While in 1914, Rajendralal Acharya had begun translating Jules Verne's novels beginning with *Around the World in Eighty Days*, in the 1930s, Kuladaranjan Ray had translated Jules Verne's *Mysterious Island* and Sukumar Ray was inspired by Conan Doyle in writing *Heshoram Hushiarer Diary*. Other attempts at indigenous production of fantastic stories had begun much earlier. Hemlal Dutta's "Rahashya", a scientific fantasy was published in 1882 in an illustrated periodical *Shachitra Bijyan Darpan* in two episodes, edited by Jogendra Chandra Shadhu. Shortlisted later in 1897 for the 'Kuntalin Purashkar', Jagadish Chandra Bose's *Palatak Tufan* was published in 1886. Jagadananda Roy had published *Shukrey Bhraman* in 1914. He claimed in the preface that the work had been written twenty-two years before its publication.[31]

The world of fantasy in print literature was not very far removed from the desire to soar high and above the mundane. The coming together of new innovations and experiments in flying and the near-realisation of the desire to defy gravity, in the context of the modern, invoked a new representational space similar to the one invoked by the camera in the realm of the visual. Obviously enough, participation in such real attempts to fly was limited to a handful compared to thousands who had flocked to make their exact resemblances in the studios. Calcutta, the usual seat of the most unusual of happenings, witnessed the ascent of Ramchandra Chattopadhyay in his own balloon named 'City of Calcutta' on 4th June, 1889. Even before that, D Robertson (1836) and a few other Englishmen tried flying in a balloon over Calcutta. But Chattopadhyay's attempt caught the popular imagination because of his Bengali identity. The *Indo-European Correspondence* wrote about this '*baboo*-balloonist': ...the Bengali *baboo*, overcoming his natural timidity, penetrated cloud-land... we would hardly have expected to find a Bengali *baboo* taking to ballooning, of all things, in the quest for fame and fortune.[32] Flying had even caught the imagination of Kalighat painters. In 1890, *Balloon e Bangali Bibi*, a Bot-tala farce by Rajkrishna Roy, referred to Ramchandra Chattopadhyay's flight while describing the whimsies of a Bengali wife pestering her husband for a balloon on which she could enjoy a fantastic ride.[33] Manindralal Bose's 1930 novel, *Ajoykumar*, serialised in the children's periodical *Mouchaak*, describes Ajoykumar's feats while travelling from Bengal to Germany and realising his dream of flying an

[31] For a detailed account see Robin Bal. *Banglay Bijyan Charcha.* (Kolkata: Shomila Press and Publications, 1996) pp. 31-54.

[32] The Indo-European Correspondence, 17 April, 1889 & 3 December, 1890 cited in Siddhartha Ghosh. *Kaler Sahar Kolkata* (Kolkata: Ananda Publishers Private Limited, 1991) pp. 92-109.

[33] Hardik Brata Biswas ed. *Prahashaner Kalikaler Bangamahila. 1860-1909.* (Kolkata: Charchapada Publication Private Limited, 2011) pp. 459-468.

aircraft.[34] These examples, although not exhaustive, are representative of contemporary investments in popular imagination.

The photographic studio, with its techniques, potentially had the ability to transform couples, depicting them against backgrounds which reflected a spatialised imagination of romance within limits of social sanction. At the same time, photographic performance acted as agency to mediate history into the present through the meaning of 'signifying properties', where the subjects become 'social actors'. However, the couples in studio photographs are constructed within the realm of an imagination which is inter-connected with the disposition of habitus through Social Science. They are not the way they appear in everyday life but, on the contrary, reproduce the cultural notions of the desired image of oneself. Here, photographs are not functioning as a reliable natural sign but they accentuate, through the multiplication of appearances, a problem of reliability inherent in natural appearance itself.[35]

Although Judith Butler's formulations on performance and performativity are not specific to the area of photographic representation, her critique can be used in this context, with obvious reservations, as she deals largely with the problem of identity politics and processes of gendering. Butler shows that identity categories are fictional products in the sense that they do not pre-exist the regimes of power/knowledge but are performative products of them.[36] They become performative in the sense that the categories themselves produce the identity they are deemed to be simply representing. Identity becomes a matter of social and political regulation rather than any sort of innate property of the individual.[37] Here, to peek into the sociology of the camera, the differences between 'performance' and the 'performative' should be taken into account. Performance of gender, according to Butler, produces a sense of an essence, which is a repetition of sustained social performances which create the reality of gender, but which, as Jagger points out, are not separable from agents or actors, preceding the performances, as in a theatrical model. What seems to be an apparent free choice to 'choose' the role one wishes to play, like in theatre, is not so much of a choice. The

---

[34] Ashok Sen ed. *Advencharer Galpa*. (Kolkata: Sishu Sahitya Samsad, 2000) pp.1-40.

[35] Udaya Kumar. "Seeing and Reading: The Early Malayalam Novel and Some Questions of Visibilities" in Early Novels in India. Ed. Meenakshi Mukherjee. (New Delhi: Sahitya Akademy, 2002) pp. 161-192.

[36] Gill Jagger. Judith Butler: Sexual Politics, Social Change and the Power of the Performative. (Oxon: Routledge, 2008) 17.

[37] Jagger. 20.

'doer' is produced in and by the act.[38] She does not stand outside the act itself, or before it, or in a position to reflect. A phenomenology of the body of the subject is necessary to such an understanding of identity and performance which are not matters of naive volition. The performance before the camera, the neatly composed couple, as it evolved in photographs since the late nineteenth century in Bengal, although at times liberatory and exceptionally framed, generally iterates the cultural and religious values of the time and region. The value of a couple photograph that is not scrutinised and appreciated by relatives and other members of the family becomes less. If not displayed, looked at, cited and iterated, that moment of exception and individuality inside the studio would prove to be something in itself that largely won't produce the effect that a couple photograph is expected to do through corporeal styles celebrating the couple form.[39]

---

[38] Most of the images I have faced of women and groups inside family albums are conditioned by durations of time exposure whose aesthetic structure was initially practiced in the nineteenth century portrait tradition and which relied on having the sitters face the camera and remain immobile for long durations of time. Benjamin concluded that direct look of sitters encapsulated them in their cocoon while the stillness that was required of them by long exposure was felt in the general impression of silence they exuded. "The procedure itself", wrote Benjamin, "caused the models to lie, not out of the instant, but into it; during the long exposure they grew, as it were, into the image."

[39] When I began to look into family albums and started talking to women about their reflections on albums and photographs, the structured mode of a closed and moulded representation began to break down fast. Women's interpretations of photographs were multiple and numerous. It varied with kinds of socialization. While I was expecting more answers from the point of view of stereotyping of women in family albums, the respondents were more eager to respond and talk about their attachments to particular snapshots. They as a general pattern talked about particular photographs as sources of memory, lost possibilities in childhood and photographs that they are not ready to submit to any determined structures. Points of dress, make-up, postures, even photographs taken during illness resurface in the minds of the respondents. One of the interviewees mentioned about a set of photographs taken during a near-fatal illness after childbirth. She told more than once the she does not like the photographs but cannot exclude them from the album. It remains as part of her essence. She goes back frequently to that particular set of photographs. Locating such experiences inside dominant genres of visualization is problematic. These are points where certain presumptions about visualization in domestic photography starts falling apart. Women (and also men) constantly reformulate meanings and positioning of photographs. Patterns of selection of 'favourite photographs' change with passage of time. Same photographs are liable to be liked and disliked with time as happens with social

Family photographs strive for unity, stability, dignity and character. Following Hirsch's argument, it can be stated that the family album is a more or less superficial psychoanalysis. We project onto the photographs and subtract from them too. Photographs in which women look 'optimistic' are liked best in albums by other members of the family and by kin. This poignancy is vested in the figure of women as a site of exchange of cultural value. Hirsch argues that the family photo has been prostituted into politics. 'He is a family man' is always a note of praise. No one as yet used 'a family woman' as a character reference. The identities of the family and the woman are exchangeable. Therefore the reference is not an extraneous one, but implicit within the body of the woman displayed in the album.

Idealised images and reality are in constant negotiation and argument in photographs, especially in family albums. The family is constituted through the iconography of significant occasions, through styles of dressing, appropriate demeanour, pose and positioning. Wedding photographs, or those taken shortly after a special event, became particularly popular and an ordeal for the bride, who is perhaps just out of *pardah*, or for the child bride. In the twentieth century, there was a certain relaxation and informality in poses of conjugal couples. In group photographs, the positioning of the patriarch as well as of others often reflected hierarchies within families.[40]

Such questions of identity and performance overlap regarding this genre of photographs.

On one hand, it is an attestation of the newly achieved status of the subject while on the other, it helps to blur, even if temporarily, the stricter everyday demands of marriage. With the nationalist movement in full swing, the 1930s and the 40s in Bengal's history were rather eventful decades. Events like the Declaration of the Independence of India or 'Purna Swaraj' on 26 January in 1930, police firing on striking *garoans* in Kolkata leading to five deaths, the Chattagram armoury raids, Hijli jail killings, strikes by labour unions, an increased presence of the *Hindu Maha Sabha* in the region, communal tension in various parts of Bengal, the famine of 1943 – all had their impact on different cultural and religious communities in Kolkata and other parts of Bengal. The middle-class *bhadrolok* and the *moddhobitto* family identity in the fields of representation do not seem to have been informed by such happenings on the ground. The *bhadrolok* imagination of conjugality and romance along

---

relations. Memory work vis-à-vis photographs are problematic and experiential readings are at times baffling.

[40] Karlekar, Intro. XV.

with a distinct way of life had been firmly put on a pedestal in the early decades of the twentieth century: These were characterised by sentiments of personal and familial attachment, a code of public and societal conduct, and a well-laid system of values.[41]

The family album and its associated photographic cultures, throughout their evolution in Bengal sought to construct a stable and happy space for the family with the couple form being its staple foundation. Given such highly complex times and histories, the studio, even if as misrecognised performative, offered the Bengali conjugal couple a fantastic life of their own. To quote Bazin:

> No one believes any longer in the ontological identity of the model and image, but all are agreed that the image helps us to remember the subject and to preserve him from a second spiritual death. Today the making of images no longer shares an anthropocentric, utilitarian purpose.[42]

Works Cited

Ashcroft, Bill, Gareth Griffiths et al. 1998. *Key Concepts in Post-Colonial Studies*. London: Routledge. Print.

Bal, Robin (1996). *Banglay Bijyan Charcha.* Kolkata: Shomila Press and Publications, 2006. Print.

Barthes, Roland. *Camera Lucida: Reflections on Photography*. London: Vintage, 2000. Print.

Baudrillard, Jean. *L'echange symbolique et la mort.* Paris, n.p.1976. Print.

Bazin, André (1967). *What is Cinema?* translated H. Gray. California: U of California P, 1992. Print.

Berman, Marshall. *All that is Solid Melts into Air*. New York: Penguin Books, 1982. Print.

Biswas, Hardik Brata, ed., *Prahashaner Kalikaler Bangamahila. 1860-1909*. Kolkata: Charchapada Publication Private Limited, 2011. Print.

Butler, Judith. *Bodies that Matter: On the Discursive Limits of "Sex"*. New York: Routledge, 1993. Print.

Chakraborti, Sambudhha. "Conjugal Relations in Early Nineteenth-Century Bengal: Conditions Before the Growth of a Private Relationship". *Women in History* Ed. Anuradha Chnada, Mahua Sarkar et al. Kolkata: Progressive Publishers, 2003. Print.

---

[41] Amit Kumar Gupta. Crises and Creativities: Middle-Class Bhadralok in Bengal c. 1939-52. (New Delhi: Orient Blackswan, 2009) p.7.

[42] Bazin, 10.

Chatterjee, Partha. *The Partha Chatterjee Omnibus*. New Delhi: Oxford U P, 2006. Print.

Crary, Jonathan. *Techniques of the Observer: On Vision and Modernity in the Nineteenth Century.* Massachusetts: MIT Press and October Book, 1992. Print.

Foucault, Michel. *Power / Knowledge: Selected Interviews and Other Writings 1972-1977*. New York: Pantheon Books, 1980. Print.

Ghosh, Siddhartha. *Chhobi Tola: Bangalir Photography Charcha*. Kolkata: Annanda Publishers Pvt. Ltd., 1988. Print.

Goffman, Erving (1959). *The Presentation of the Self in Everyday Life*. London: Penguin Books, 1990. Print.

Gupta, Amit Kumar. *Crises and Creativities: Middle-Class Bhadralok in Bengal c. 1939-52*. New Delhi: Orient Blackswan, 2009. Print.

Hajra, Niradbaran. *Kolkata: Itihaser Dinlipi*. Kolkata: Pashchimbanga Rajya Pustak Parshad, 1995. Print.

Hall, Stuart. *"Cultural Identity and Diaspora"*, in Identity, Community, Culture, Difference, ed. Jonathan Rudeford. London: Lawrence and Wishart, 1990. Print.

Jagger, Gill. *Judith Butler: Sexual Politics, Social Change and the Power of the Performative*. Oxon: Routledge, 2008. Print.

Karlekar, Malavika. *Re-visioning the Past: Early Photography in Bengal 1875-1915*. New Delhi: Oxford U P, 2005. Print.

—. *Visualizing Indian Women 1875-1947.* New Delhi: Oxford U P, 2006. Print.

Keopnick, Lutz. "Photographs and Memories". *South Central Review*. 21.1 (2004). Print.

Kumar, Udaya. "Seeing and Reading: The Early Malayalam Novel and Some Questions of Visibilities" in *Early Novels in India*. Ed. Meenakshi Mukherjee. New Delhi: Sahitya Akademi, 2002. Print.

Majumdar, Rochona. *Marriage and Modernity: Family Values in Colonial Bengal*. New Delhi: Oxford U P, 2009. Print.

Parayil, Sujith Kumar. "Photography and Colonial Modernity in Keralam" in *Space, Sexuality and Postcolonial Studies.* Ed. Manas Ray. Kolkata: Centre for Studies in Social Sciences, 2003. Print.

Pinney, Chirstopher. *Camera Indica: The Social Life of Indian Photographs*. London: Reaktion Books, 1997. Print.

Prakash, Gyan. *Another Reason: Science and the Imagination of Modern India.* New Delhi: Oxford U P, 2000. Print.

Raychaudhuri, Tapan. "Love in a Colonial Climate: Marriage, Sex and Romance in Nineteenth-Century Bengal". *Modern Asian Studies* (2000). 34. Print.

Said, Edward. *Orientalism*. New York: Vintage, 1978. Print.
Sen, Ashok, ed., *Advencharer Galpa*. Kolkata: Sishu Sahitya Samsad, 2000. Print.
Sontag, Susan. *On Photography*.New York: Farrar, Straus and Gironx, 1977. Print.
Tagg, John. *The Burden of Representation: Essays on Photographies and Histories*. London: Macmillan, 1988. Print.

CHAPTER TEN

DRAUPADI AND US

SAOLI MITRA

About twenty-eight years ago, I had a feeling that what was happening around me was not right. I felt the need to express myself. That urge inspired me to write a play. I had never imagined I would be able to do so. It came to be known as *Naathavati Anaathavat.* It was based on the *Mahabharata*. It looked at the then society from society's point of view. The protagonist of the play was Draupadi, the peerless beauty and hapless queen. It was written in the traditional storytelling form. A narrator, *kathakaar* or *kathakthakrun* narrated the story. Another play was written in 1980-90 in the same style. It was called *Katha Amritasamaan* and based on the epic again. *Katha Amritasamaan* literally means 'Words like Nectar'. It describes the annihilation of the era.

Today, I shall begin with the story of Draupadi and conclude with the story of the destruction of an era. I hope it will not sound too ambitious if I mention that with the writing of the play, a narrator or a *kathaktbakni,* was born in modern Indian theatre. In the play, a lone *kathakaar* tells a story, relates it to the present day and comments on the situations described. Thus it is easier for us to relate to Draupadi, the queen of the Kuru clan of Dwapar Era.

Despite what has been said for a long time, a query emerges – why this urge to write a play? It haunts you. Gradually you find the answer. Socio-political and socio-economic factors prompt you to say something, to discover your position in the society you live in. The cultivation of your creative self thus begins within yourself almost unknowingly! It is sad to find that the position of women in the 21st century is no better than it was decades back. Is it really different from the situation of Draupadi of the *Mahabharata*? Unfortunately the answer is 'no'.

It should be mentioned that an art form does not merely depict reality. It converts physical reality into a metaphysical truth. Art aspires to express truth and not mere reality. It speaks of reality and at the same time, expresses rare human experiences to convey the complexities of life. It is interesting how we have a situational compulsion to get on with our lives and yet decide what is right and what is wrong. It indicates that our existence is inevitably related to a sense of morality. I strongly believe that human civilization cannot exist without this sense of morality. The complexities of life emerge through the questions raised before us. The *Mahabharata* gives us enough scope to venture into a quest for truth. It depicts human lives with unfathomable depth and an enormous wealth of wisdom.

I narrate the story of the queen not just to give an account of her life. We realise the agony of the Yuga through the life of the most beautiful, intellectual and educated woman of her time – a woman who suffers profound agony over and over again. We also compare Dwapar Yuga to the present time in both my plays. I, as the *kathakaar*, quote Buddhadev Basu, a well-known poet and intellectual of Bengal, and, in particular, his analysis of the epic. I also incorporate Smt Iravati Karve's interpretation (Karve is a renowned anthropologist and an authority on the *Mahabharata*). The folk form assimilates their works effortlessly.

I. Draupadi

The mere fact that Draupadi is forced into marriage to all five Pandavas reveals the intricacy of the tale! It reveals the social structure of the day. It reveals the patriarchal structure of the play. The decision is taken by the guardians of society. And that, of course, is a political decision. The mother, Queen Kunti, looks at the girl in her bridal attire and feels that each of her five sons are attracted to her beauty. It is almost immediately after the *jatugriha* episode. The Pandavas are still disguised as Brahmins and are yet to restore their right. If they fight among themselves at this juncture, their future is bound to be destroyed. Poet Vyasa and the great Krishna understand the situation too. So they put forward several examples from the *puranas* before Drupada and prove that it is nothing abnormal (for a woman to be married to five men simultaneously). Moreover, Arjuna is the third among the Pandavas and the eldest, Yudhisthira, is still unmarried. Hence, it is necessary to get all of them married to the princess. Isn't it a terrible blow for the beloved young daughter of Drupada? But everybody accepts the idea and she is married separately to each of them. They further decide she has to spend one year in turn with each of them. This, too, should start from the eldest.

We remember how a sensible, cultured and exquisitely beautiful Draupadi falls in love with a poor yet capable and dark yet handsome young Brahmin and gladly accepts him as her husband. She chooses

Arjuna in the *swamvar sabha* not knowing his true identity. The young girl is full of joy and happiness!

> Bonopathe chole oi veer Phalguni
> Saathe chole brishtochitte Drupadanandini

But where is she herself? She is forced to put up with the humiliation of living with five husbands for the rest of her life. The *Mahabharata*, however, says that she accepts it 'gracefully'. Draupadi thus endures insults – sexual and otherwise – all her life. The complexities she faces and the ways in which she tries to retain her integrity are incomparable. Draupadi, the mother of five sons, is used as a pawn in the notorious dice game. Yudhisthira, king of Indraprastha and son of Dharma, does not hesitate to commit the immoral act of pledging his wife as a prize in the game. The *kathakaar* narrates in the play:

> Look! There! The dice is thrown. Shakuni has won. For shame! What obscure hilarity in the assembly! Karna, Duryodhan, Duhshashan, Shakuni – what brutal lust on their faces! Meanwhile, the Pandavas sit speechless. And the wise elders? What about them? They, too, have been robbed of speech! Good sirs, there are times when the wise and the learned keep mum, while the weak go on being tortured endlessly! Endlessly!
> *Shukhi borro shukhi chhilo pandavapriya*
> *Shukhi chillo jobe chhilo drupadatonoya*
> (She was happy when she had been Drupada's daughter).

There is great exaltation at court. Kama says to Duryodhan, "Friends, can the Pandavas still occupy the royal seats? They are now slaves, after all!" As soon as they hear this, the defeated Pandavas forsake their rich robes and ornaments as well as their seats. At this, Duryodhan and his courtiers break into obscene cackles.

What happens in court after that? Draupadi is dragged there by her hair by Dushashan, the second Kaurava. The matter does not end there. The *kathakaar*, donning the persona of Draupadi, observes: "Bhishma, Vidura, Dhritarashtra, Dronacharya – have they no feeling? I am a daughter-in-law of the Kuru clan. I've been dragged into this assembly and nobody condemns it! Please say something!"

Old Bhishma, the head o the clan, says, as if judging the situation with his wisdom, "Look, Yudhisthira is the most learned in precepts. And he himself says that he has been won over. Therefore, you see, it is very difficult to…" – Interesting indeed! But we are startled anew when we find Karna suggesting they disrobe the woman in court! The *kathakaar* of the play describes the situation in Kuru's court:

> … And then it is Karna, good sirs, Karna, the brave and benevolent, the tragic hero of the modern imagination, the victim of fate, it is Karna who says, "Yudhisthira has been won with all his wealth and property and Draupadi is merely a part of it. I find it quite intolerable to split hairs over this." The great and gallant Karna then says, "Duhshashana, go and strip the woman in front of the court. And let us watch. Come on, come on, Duhshashan!

Duhshashan approaches her and tries to strip the sari off a woman who is menstruating! The woman is helpless and is unable to flee the court in that condition. She has only one piece of cloth to cover herself as per the ritual of the day. Pale with shame and fear, she begs Duhshashan to stop. But the Kauravas enjoy the situation enormously. It is fun, I believe, to watch a woman who is as trapped as an animal. So they break into shrieks of laughter.

Finally, Draupadi leaves everything behind and walks into exile for 12 years with her husbands. There too, she faces disgrace and sexual abuse. Most of the time, Arjuna is away from her. He has his duties towards his brothers and towards the lost kingdom. Once Draupadi meets him in solitude and tells him, "Partha! Your brothers may find consolation in discussing your feats. But when you are away for too long, I find no joy in anything – neither wealth, nor pleasure, not even in living." Does this not reveal her forlorn state of mind? Her love for Arjuna? Even if we ignore the other instances of abuse in the forest, can we forget the episode of Kichaka, the brother-in-law of King Virata? Kichaka kicked her in the face in court. But he is supported by the king!

It is not hard to realise, when you go through the play, that it is a critique that connects women of the present age to Draupadi. Women who suffer similar agony and humiliation! We know what happened to Tapasi Malik in Singur and to several women in Nandigram and Lalgarh in our own state! Most of us endure disgraces of different kinds in society.

Draupadi, in my play, asks Krishna a vital question. Krishna, the lone representative of the Pandavas, is about to go to Hastinapur to attend a so-called 'peace conference'. Draupadi comes up to Krishna and reminds him of the indignity she once faced in the Kuru court. She finds it amazing that sitting in the same court, her friend and guide, is about to talk about 'trust'! To talk about 'peace'! She asks:

> If I forget the humiliation inflicted on me, dear friend, will it usher in a *Dharmarajya*, a rule of virtue? Can you promise that in the future, no woman will ever be persecuted and humiliated as I was? Will my forgiveness usher in that heavenly state?

Was Krishna, the powerful political leader, able to answer this question? No, He keeps mum. Draupadi earnestly says, "If you promise me this, Krishna, I tell you, I would forget everything… everything!"

In this respect, we must remember that Draupadi is quite different from Sita, who is considered an icon of ideal Indian womanhood. Draupadi protests injustice. Draupadi argues. She argues even before the court. She blames Yudhisthira for bringing calamity to the entire family. She, quite sarcastically, reproaches Arjun for his involvement with other women. So, it is not difficult to discover the anguish of a modern woman in her. This element probably prompted me to choose Draupadi as the central character of my play.

She loses her sons to the war. She becomes the queen of a kingdom where there are no young faces. A kingdom that belongs to widows and childless mothers! In her final journey with her husbands, Draupadi is the first one to fall. She learns it is because she secretly loves Arjun the most. It is considered her sin!

II. Society and Art Form

Somehow I feel that in a male-dominated society, especially one that loses its sense of morality, women have to undergo undesirable experiences. Unfortunately, we belong to an insensitive civilization that disregards humanity itself.

I feel theatre shows human emotions as well as faces. Both actors and viewers are keen to know the characters on stage with all their shades and multifaceted characteristics. They look for human beings of flesh and blood. That makes us work hard so that we can create the image as clearly as possible to convey it to our audience. By enacting them, we make the characters transparent to the actors, and, I hope, also to viewers. We show what we are and what we should be.

III. Human existence

Now we come to the question of human existence. Isn't it true that at this juncture, humanity humiliates itself? Isn't it a time when every moment, the spirit of an individual is being insulted? Unfortunately, we belong to a time when quantity gets more importance than quality. But can we ever forget that an individual always remains the central point of a civilization? Can we deny that a balance is constantly needed between an individual and society? We may further add that every human being has two selves that work simultaneously within him. One is his individual self and the other, his social self. A good society helps in maintaining the equilibrium between the two selves.

As I mentioned before, theatre is not for mere depiction of a story. One has to express one's commentary on one's time. Artistes express their emotions and their commentaries through theatre, or any art form for that matter. It is also necessary to raise relevant questions about civilization in general. We believe that a civilized society loses its stature and its dignity if its members, especially intellectuals and the artists, are not concerned enough about the society they live in.

IV. Our Time

Isn't it strange that while we call the earth a global village, we allocate the highest budget for buying weapons to kill people? Shouldn't we question the justification behind maintaining such double standards? Shouldn't we talk about the innumerable scams of political leaders? Shouldn't we question storing of illegal arms by goons? Don't we need to ask why the mischief-makers are protected by the administration itself? We remember what happened in Iraq over the last four or five years. How do we react? Do we remember what has been happening in Kashmir or the situation in our neighbouring countries, Pakisthan or Myanmar? In Afghanisthan? In Libya? Don't we feel that something is very, very wrong with this civilization and its norms? We pay a heavy price for the atom bomb that targeted people of Hiroshima and Nagasaki. Shouldn't we remember that soldiers who returned to America after the gulf war in the 90's gave birth to forty thousand deformed children? The children and their mothers had no connection with the war whatsoever. This is the reality we are living in. Greed for power creates a materialistic world that is only concerned about money and wealth. Human society of the 21st century builds shopping malls, highrises and multiplexes. This is a society that continuously murmurs – 'Buy, buy and buy'.

This is the culture of modern society, especially in the metros, in cities and in industrial townships. It is most unfortunate that this culture has been thrust upon the common people of our country. It disregards those very characteristics that distinguish man from other animals – the sense of values, the sense of justice and, of course, the sense of morality. This precipitates the undesirable end of human existence.

V. *Raktakarabi*

Rabindranath Tagore wrote his famous play, *Raktakarabi* (Red Oleanders), in 1924. He warns mankind about these fearful consequences. A poet and a philosopher, he cautions mankind against greed, against regimentation of society and against humiliation of the individual self. He speaks of agrarian culture and compares it to the culture of the industrialised sector. He never bothered to care. We are quite contented to observe 25th Baisakh and 22nd Srabon (the birth and death anniversaries of the great thinker). I would like to add that we should refrain from coming to a decision that the poet is against industrialisation altogether. He only recommends that we be conscious of our roots and our culture. Nandini, a protagonist in *Raktakarabi*, says, "The earth gives us her wealth with affection, with love; but when you grab something against her will, you invite a bad omen. You induce the curse of a blind fiend. It prompts killings and snatchings in society." She warns the scientist (in Bengali, the character is identified as Raja who lives behind a net) to come forward to use his might for Mother Earth and for the people who love Mother Earth. Raja fails to understand what she means. It is too late to save society from bloodshed when he realises the situation.

VI. Culture

We utter an important word – culture. What is culture anyway? We must remember that culture does not only mean a bit of theatre, a bit of music, a painting exhibition or a dance recital – culture is actually a life pattern, a life a particular community intends to live. The life depends upon their means of livelihood, the natural environment they live in and the way they accept moments of misery and joy! Culture means the whole cycle. Culture generates humanity, fellow feeling, love for nature, self respect and also a sense of morality. The cycle, if ignored, creates imbalance in human society. Eventually, it endangers the existence of civilization itself. And perhaps that is what is happening today!

Theatre cannot solve these problems. Neither does it have the strength to change the world. It can merely raise some vital questions for our existence. We feel that is the need of the hour. We, each of us, should try to protest the inertia that takes the lead role now for the so-called progress of human society and pushes us into complete blindness.

VII. Conclusion

Now I am tempted to mention something about *Katha Amritsaman.* The play concludes by depicting the last days of three very old people – Dritarashtra, Gandhari and Kunti. Once powerful, they now live in the forest and fight poverty. They neither have a present nor a future. They only look forward to reaching the end of their lives. In the evenings, the three of them sit by the riverside. On one such day, the blind Dritarashtra smells fire. At once he guesses it is a forest fire. Kunti is the only person among them who is able to see. She looks at the fire, at the intense smoke, as it moves towards them. She even feels its heat. Suddenly she remembers her husband Pandu. She believes she will meet her beloved husband after her death. The sooner the death comes, the better. Yes! God has at last been kind enough take her into his arms! Death is approaching her, she feels happy! She too has suffered agonising pain in life. How does she feel when she recollects Pandu, her husband, lying dead on the lap of his younger queen Madri? How does she feel when she remembers the attempt of the Kurus to burn her children alive? How does she feel when she remembers Draupadi's insult in front of the entire court? She loved and still loves Pandu! She is sure now that the time has come to meet him. She suggests to the elders that instead of trying to escape from the fire, they should surrender to it. They all agree. Then the three very old and weak people walk towards the fire. They prefer to die with dignity than to live in disgraceful circumstances. Eventually, the forest fire engulfs them. They are burned alive.

The *kathakaar* announces the end of the era. She recites a *shloka* from the *Mahabharata* and prays for a better world – a world that allows people to live with peace and does not kill them. And we remember Vyasa:

… Atikranta sukhah kaalah
Poryupasthita daarunah
Shwah shwah paapishtha divasah
Pruthivi gata yauvana…

Gone are the days of happiness
The earth has lost her youth.
The world faces
The menace of destruction.
Lost is the meaning of justice.
Sins increase a hundred-fold.
The world is full of wails!

We pray,

Let there be light!
Let all sins be wiped out!
Let the sky be clean!
Let the wind be pure!
Let this earth be free of sorrow!
Let humanity live!
Let the leader emerge!

We quietly wait for mankind to rise against Evil.

We, from the darkness around us, pray for Light. Light!

Notes:

1. The translation of the prayer pronounced in my play:

Oh Eternity, today at this juncture of great change,
For our children's sake at least, let there be justice, morality, truth and humanity, established on this earth once again. For our children's sake at least, make this earth a beautiful good place.
Let the darkness be wiped out! The darkness that has engulfed our earth
O Mahakaal, make the earth beautiful in every possible way for our progency, for the generations to come in future! …

2. I wrote *Katha Amritasamaan* in 1990 to depict the agony of the era, to discover the reasons behind the annihilation of the *yuga.* Both the plays have been translated into English and been published by *Stree* publication.
3. All textual quotations are from my play on Draupadi, *Naathavati Anaathavat* written in 1983. The book has been translated into English as *Five Lords, Yet None A Protector* and *Timeless Tales.* It is a collection of plays and two essays called *An Imaginary Conversation with My Audience* and *Krishna: An Enigma*

Glossary

Arjun – The third of the Pandavas, the sons of Pandu, who, with Krishna, is considered the hero of the Hindu epic, *Mahabharata*. He plays the role of a listener in the Hindu scripture, the *Bhagavad Gita*, which is a philosophical conversation between Arjuna and Krishna. Arjuna, was considered the finest archer and a peerless warrior of his time.

Bhishma was the eighth son of Kuru king Shantanu,[1] who was blessed with a life that would be as long as he wished and had sworn to serve the ruling Kuru king.[2] He was one of the most prominent characters of the great Indian epic. He was the grand uncle of both the Pandavas and the Kauravas.

Bhima is the second of the Pandava brothers. He is one of the central characters and symbolises great strength.

Draupadi – Daughter of King Drupada of Panchāla and the wife of the five Pandavas.

Drona or **Dronacharya** was the royal guru to Kauravas and Pandavas. He was a master of advanced military arts, including the Devāstras. Arjuna was his favorite student.

Dritarashtra – King of Hastinapur at the time of the Kurukshetra War, the epic's climactic event. He was born the son of Vichitravirya's first wife Ambika, and was fathered by Vyasa. He was blind from birth and became father to a hundred children borne by his wife Gandhari. These children came to be known as the Kauravas.

Duryodhana, a name which means 'difficult to fight with' (his real name was Suyodhana) is the eldest son of the blind king Dhritarashtra by Queen Gandhari and the eldest of the one hundred Kaurava brothers. He was the cousin and the chief antagonist of the Pandavas.

Duhshasana was the second son of the blind king Dhritarashtra and Gandhari in the epic and the younger brother of Duryodhana.

Dvapara Yuga or **Dwapara Yuga** is the third of four yugas, or ages, described in Hindu scriptures. This yuga comes after Treta Yuga and is followed by Kali Yuga.

Gāndhārī was an incarnation of Mati, as the daughter of Subala, the king of Gandhara, Gandhari's marriage was arranged to Dritarashtra, the eldest prince of the Kuru kingdom. Gandhari voluntarily blindfolded herself throughout her married life. Her husband Dhritarashtra was born blind and on meeting him and realising this, she decided to share the pain of her blind husband.

Jatugriha parva – After the deaths of their mother (Madri) and father (Pandu), the Pandavas and their mother Kunti return to the palace of

Hastinapur. Yudhisthira is made Crown Prince by Dritarashtra under considerable pressure from his kingdom. Dritarashtra wanted his own son Duryodhana to become king and lets his ambition get in the way of preserving justice. Shakuni, Duryodhana and Dusasana plot to get rid of the Pandavas. Shakuni calls the architect Purochana to build a palace out of flammable materials like lac and ghee. He then arranges for the Pandavas and the Queen Mother Kunti to stay there, with the intention of setting it alight. However, the Pandavas are warned by their wise uncle, Vidura, who sends them a miner to dig a tunnel. They are able to escape to safety and go into hiding. At Hastinapur, the Pandavas and Kunti are presumed dead.

Kunti was the daughter of Shurasena and a Yadava and the wife of King Pandu of Hastinapur. Before Kunti married Pandu, she bore Karna to Surya, the solar deity. She later married Pandu and bore Yudhisthira, Bhima and Arjuna.

Karna is one of the central characters in the epic. He was the King of Anga. Karna was one of the greatest warriors whose martial exploits are recorded in the *Mahābhārata*. Karna was the son of Surya (a solar deity) and Kunti. He was born to Kunti before her marriage with Pandu. Karna was the closest friend of Duryodhana and fought on his behalf against the Pandavas (his brothers) in the famous Kurukshetra War.

Krishna is the eighth avatar of Lord Vishnu. The earliest text to provide descriptions of Krishna is the *Mahabharata,* which depicts Krishna as an incarnation of Vishnu. Krishna is central to many of the stories in the epic. The eighteen chapters of the sixth book of the epic that constitute the *Bhagavad Gita* contain Krishna's advice to Arjuna on the battlefield.

Kichaka or Keechaka was the army commander of Matsya, the country ruled by **King Virata**. He was also the brother of Sudeshna, the queen. Malini is Draupadi's name when she is disguised as 'sairandhri' (female servant) in King Virata's palace for one year. Keechaka lusted after Draupadi but she ignored his advances. On one such occasion, when Draupadi had rebuffed his attempts to violate her honour, she was pursued by Keechaka to the throne room, where she was once again humiliated before a full assembly of courtiers, including her disguised husband Yudhishtira and King Virata himself. Neither of them protested because Keechaka wielded tremendous power within the kingdom.

Madri – The second and younger queen of King Pandu.

Mahabharata – The great Indian epic by the sage **Ved Vyasa**.

Pandu is the son of Ambalika and Rishi Veda Vyasa. He is more popularly known as the father of the Pandavas and ruler of Hastinapur.

The **Pandavas** are the five acknowledged sons of Pandu by his two wives Kunti and Madri. Their names are Yudhisthira, Bhima, Arjuna, Nakula and Sahadeva.

Tapasi Malik – On 18th December 2006, 16-year-old Tapasi Malik, the only daughter of Monoronjan Malik, a sharecropper in Singur, West Bengal was gang raped and then taken to an open pit and burnt alive to remove evidence of the rape.

Yudhisthira was the eldest son of King Pandu and Queen Kunti. He was king of Indraprastha and later of Hastinapura. For his piety, he was known as *Dharmaraja* (which may be translated as either 'righteous king' or 'king of dharma').

CHAPTER ELEVEN

MIRABAI AND INDUBALA: SPIRITUAL EMPOWERMENT REDEFINED

BIKAS CHAKRABORTI

The word 'empowerment' means an enhancement of the political, social, economic and spiritual strength of individuals and communities that seem to be marginalised. Marginalisation refers to the overt or covert trends within a society by which individuals or communities are deprived of decision-making power by the wider society and ostracised as undesirable. Thus conceived, women's empowerment necessarily presupposes the marginalised condition of women in society in general.

A very pertinent question crops up: Why are women marginalised and by whom? To borrow a Marxist explanation to the problem, Frederich Engels' *The Origin of the Family, Private Property and the State* may be cited as a brilliant exposition of the class approach to gender differences. Analysing the dialectical relationship between the development of private property and its inheritance and the structure and functions of family, Engels made a nice comment – "In the family, he is the bourgeois and the wife is the proletariat" (Engels 122). What this intends to mean is the total hegemony of the bourgeoisie over the proletariat in a class-dominated society; so is the subjugation of women by men in a male-dominated family structure having roots in the division of society into contending classes. At a certain stage of social development, 'monogamy *only for the women* (ital. original), but not for the man', was enforced as a rule in order to ensure the succession of a man's property to his descendants only. Take, for example, the case of the Hindu social organisation in which the canons of the *Manusamhita* demand wives to follow the dictats of their husbands. Moreover, the birth of a son is preferable to that of a girl since a male member will develop as a productive unit in an agricultural setting. Contrarily, the chief function assigned to women is that of reproduction and that is why emphasis was laid upon the protection of wives with utmost care and attention – *varya sarbaprayatne rakshaniya.* Words like 'sacrifice', 'care', 'love' and 'affection' are identified as synonymous with the supposed qualities of women only. What stands out is that although women bring life to earth, they live without certain basic rights in the socio-economic plane.

The moot point in the concept of women's empowerment is the ability to exercise full control over one's actions. So far as the status and role of women in our society are concerned, there have been some changes in concepts over the decades, that is, from 'welfare' in the seventies to 'development' in the eighties and then to 'empowerment' in the nineties. This process has gained some momentum with some sections of women turning increasingly self-conscious of their discrimination in several areas of family and public life. Different activist groups upholding the cause of women's empowerment have advanced different approaches to the problem. Without entering into the intricacies of these approaches, the parameters of women's empowerment as provided by the United Nations Organisation may be taken into consideration. According to the UNO, the concept of women's empowerment has five components: (1) women's sense of self-worth; (2) their right to have and determine choices; (3) their right to have access to opportunities and resources; (4) their right to have power to control their own lives, both within and outside the home; (5) their ability to influence the direction of social change to create a more just social and economic order, nationally and internationally.

My intention here is not to advance a long discourse on the theoretical aspects of women's empowerment. On the contrary, I would like to focus my attention on the gains of spiritual empowerment of two feminine personalities who, in turn, promoted the uplift of women around them. In the first case, Mirabai (1498-1547), the legendary ascetic singer of Rajasthan, will be cited. In the second case, achievements of Indubala (1898-1984) as a musical performer and actress will be discussed. The time-gap between the two personalities is four hundred years but both of them took resort to music to express and elevate themselves. The highest common factor in this case is that both Mirabai and Indubala set imitable examples to others indirectly or directly on how one can empower oneself spiritually in the face of social constraints and at the same time, imbibe others' strength for their self-empowerment.

Mirabai was born in 1498 in Merta of Rajasthan. She had a rigorous training in music from childhood and also practised religious leaning. The influence of Rao Dudaji, her grandfather, who was a devout worshipper of Vishnu, was very important in her life. Legend goes that a mendicant gifted her an idol of Krishna. Mira became so attached to it that she kept the idol with her at the time of her marriage ceremony and also took it with her to Chitor. A conflict cropped up when she declined to worship Kali, the goddess worshipped by her husband, Bhoj Raj, thereby incurring the wrath of the Rana family as a whole. Even after the premature death of Bhoj Raj in 1521, Mira continued to be a victim of political intrigues in Chitor. During her widowhood, she devoted herself wholeheartedly to the worship of Krishna and would offer hospitality to all travelling mendicants. She was unequivocal in identifying Krishna as her divine lover and husband: 'Mira's lord is Girdhar Nagar / She is the slave of his lotus feet'. Or, 'I have only Girdhar Gopal / And no one else at all'. Extremely unfavorable circumstances in both Chitor and Merta compelled her to leave Rajasthan and take refuge in Vrindabana and later in Dwarka where she breathed her last in1547.

It is important to note that Mira's refusal to consummate her marriage and her firm resistance to the chauvinistic norms of the Rana family and her simultaneous adoration of Krishna branded her as *kul-nāsi*, that is, one who tarnished the image of the clan. Parita Mukta, a researcher on Mirabai, explained the nature of the contradiction:

> The antagonism of the Sisodiyas to Mira was not merely a sectarian one of Shaivites against a Vaishnav worshipper. It was, rather, a wider battle between an individual holding on to the principle of love through Krishna – and the political authority of the Ranas based on force and might. (56)

Further, 'Mira broke the loyalty to *kul*, to prince and husband and created a new life based on love' (Mukta 66).

Mira was an ardent devotee of Krishna but never intended to be a guru, amass a band of followers or become a follower herself. But she began to have followers among the *dalits* and the peasant communities of Rajasthan nonetheless. Her association with Rohidas, who became a significant symbol of self-assertion of Chamars in Punjab and Uttar Pradesh, established a link between the *dalit* communities and Mira. Mira held Rohidas in high esteem and her acceptance of him as a guru seemed to be a jolt to the established pattern of caste relationships. In Saurashtra, Rohidas' personality appealed to the imagination of Mira *bhajan*-singers as well as non-*dalits* wishing to share the bitter experiences of untouchables. The spiritual support of Rohidas in emboldening Mira to stick to her decision to reject the Rana is very pronounced in the Saurashtrian *bhajans*. They read like this:

> Mira found a guru in Rohidas
> She bowed at his feet and sought his blessings
> Mira's Mohan (has) come to the desh of the Mertani.

Moreover, the Anjania Patels, the Kunbis and the Minas around Deogarh also felt supportive towards Mira mainly because of her voice of resistance against the Rajput feudal power and authority.

An empirical study by Parita Mukta suggests that 'One of the most remarkable features of Mira *bhakti*, as enunciated by the subordinated groups in Rajasthan and Saurashtra, is its sustained challenge to norms governing the marital tie and the marital life' (115). What is interesting to note is that *bhajan* singers have developed an image of Mira after their own personal experiences and realisations. Thus Mirabai, having a royal lineage from both her paternal and in-law's sides, became people's Mira. The *bhajans* highlight the conflict between Mira and the Rana and glorified Mira who contemptuously rejected all symbols of her conjugal relationship. *Bhajaniks* also place emphasis on Mira's own construction of her relationship with Krishna. In a *bhajan* collected from the village, Ghanti, of Udaypur, Mira is supposed to claim unequivocally, 'The Lord of Dwarka has wedded me / My hand was joined with Hari's'. Repetitive use of the words 'worship Hari' in the form of a refrain is an example of Mira's legitimisation of marriage with Krishna – one that she accepted from the core of her heart rather than the socially imposed one (Mukta, 130). When a piece of music is composed, it is likely to be shared by other singers or groups for performance. Similar is the case with this *bhajan* that, although emanating from the leather community, is sung by a number of peasant communities.

In another *bhajan* widely sung in Saurashtra and Dwarka, Mira, on the one hand, speaks without any reservation of her attachment to Krishna and, on the other hand, unveils the sufferings and hollowness of practical life. *Bhajaniks* represent Mira's version:

> The domestic life is a sour and hollow one
> One weds – and then becomes a widow
> Why should I go to his house, beloved Mohan
> I have become attached to your face

It is clear from this *bhajan* that her second point is a justification of the first, that hers is indeed 'an indestructible relationship' with Mohan. In other words, "The people's Mira continues to stand up in song against an imposed widowhood as well as an imposed marriage" (Mukta, 148-49). Mention should be made that apart from the contents, the style of singing is very indicative so far as *bhajaniks* in Saurashtra are concerned. The solo voice within the collective rendering of *bhajans* represents the pleasure and pain of Mira's personal being. The collective way of expressing the values of personal liberties Mira stood for and the projection of Mira's personal self through a solo voice are complementary to each other. In other words, the content and form of the *bhajaniks'* performance corroborate each other (Mukta, 168).

So far as the performing parts are concerned, a comparison may be drawn between the *bhajaniks'* presentations and that of *Jhanpan* songs dedicated to Manasa, the goddess of snakes. Samples of my personal collections of *Jhanpan* songs suggest that Behula appeals in a solo mournful tone to the snake charmer to bring her husband, Lakshindar, who died of snakebite, back to life. On the other side, other performers, four or five in number, represent, in chorus, the helplessness and inability of the snake charmer to respond to Behula's piteous request because he does not wish to incur Manasa's wrath. The intention of the band of singers and *Jhanpan* songs, too, is to induce people to worship Manasa who is supposed to protect people from snakebites, especially in monsoon. Choral presentation, in this case, stands for the might of Manasa while the solo voice of Behula hints at the helplessness of man against the wrath of the goddess.

To come back to Mira's case, what is significant is that Mira stands out as the symbol of 'a common lack of personal freedom'. *Bhajan*-singers, facing similar painful experiences in their personal lives, vented themselves through singing and have developed a kind of community feeling and also identified with Mira. In her concluding observation, Parita Mukta says:

> The *bhajniks* of Saurashtra and Rajasthan have given of their impulses and energies to remake Mira and through this have remade themselves… They have encapsulated a vision of Mira through whom they project their own history as well as showing the centrality of Mira in cementing alternative bonds. There is a collective, political and deeply personal bonding here. (169)

Indubala, who was known as an empress in the world of music of Bengal and also became a legend in her lifetime (1899-1984), was born at a time when more than forty years had elapsed since India was placed directly under the British rule following the Sepoy Mutiny of 1857.Under colonial rule, the marginalisation and subjugation of one section of people by another acquired a new dimension in India because the world of ideas, in this case, was also dominated by colonial masters who thought it their divinely ordained responsibility to arrest the moral decline of Indians and civilize them. The class of neo-literate Indians, nurtured by the system of education set up by Christian missionaries, developed a kind of abhorrence and neglect towards certain aspects the performing arts associated with *tawaifs* and prostitutes enjoying the patronage of nawabs and zamindars, and also a host of folk elements of the indigenous culture.

Many protest movements were organised and a number of associations sprouted all over India with the goal of cleansing Indian society. The move to abolish *nautch* parties took off in Madras around 1892-93, followed later by official declaration of all '*nautch* girls' mere prostitutes. The government of the Maharaja of Mysore followed the trail to ban the *devdasi* system and public dance performances as early as 1909. The anti-*nautch* movement spread to other parts of India too, with the Punjab Purity Association of Lahore, the Social Service League of Bombay and a host of similar organisations in Northern India being started with the same objective in view.

These so-called cleansing moves dealt a death blow to the profession of performing women. No route other than prostitution was left to those who could survive the onslaught. History is replete with evidences of *tawaifs* in Northern India adopting a strategic change in their performance delivery by emphasising the musical part and, at the same time, forsaking the dance element. Even the surname 'Bai' became synonymous with infamy while 'Devi' could earn some social status. Reba Muhuri, the renowned *thumri* performer, in her book *Thumri O Baiji*, referred to an incident in which Badi Motibai (1885-1977), an acclaimed vocalist of Benaras, took the host of a musical programme to task for his contemptuous behaviour towards her. The situation worsened when the host and his family members refused to dine with the performer. The author reacted sharply and they were compelled to have their meal together (Muhuri, 60-1). It is important to mention that the impact of this puritan movement was not uniform in nature. Vikram Sampath, the biographer of Gauhar Jaan, points out:

> The anti-*nautch* movement made little difference to the high class *baijis* like Gauhar. The fact that Gauhar took the music out of the *kothas* and into the more largely accessible format of the gramophone discs, which could be heard and appreciated by a large cross section of society, was in itself the first major step towards liberating the arts. Performance standards and protocol as also the life-styles of performing musicians, especially women, changed dramatically with this recording revolution. In spite of such developments, the *tawaif* community was denied any recognition by the then social, political and administrative quarters. (187-88)

There is no reason to believe that this contempt for the *baijis* was not protested. Amiya Nath Sanyal, a connoisseur of Indian classical music and also a musicologist par excellence, presented a cogent argument:

> People consider *baijis* to be debauches. Because they had to entertain a host of men in their lives – it is just a profession. They earn money for their survival by enchanting many by music... The reason is economic. What is the cause of treating them with contempt on grounds of profession only?

Further, "In my opinion, each and every great *baiji* is a *gandharbee*. In fact, it is our misfortune that we could not pay them the respect that they deserved". (Muhuri, 13-5)

Indubala was born at Amritsar on November 1899. Her parents were Professor Matilal Bose, who was the proprietor of Great Bengal Circus, and Rajbala, who was a player in Bose's circus. The honeymoon in their conjugal life did not last long and they separated when Indubala was only three years old. In her reminiscences, Indubala says:

> My mother became totally companionless when my father deserted. She had none to depend upon and I seemed to be an added liability for her. Yet she dared stand up. The way she adopted to survive and to save me in the hard struggle of life was neither very pleasant nor respectable. Still one would of course forgive my mother if one can well appreciate her helplessness in those days when women lacked any respectable means to hold their heads high. My mother with her three-year-old child took shelter in Kolkata's Rambagan area, then known as Rupogachhi'. (88)

It may be noted that this locality was, and is, one of the major red light areas in Kolkata. Rajbala concentrated on music as a means of livelihood and also took part, as an actor, in contemporary theatres.

Indubala had her initiation into music under the tutorship of her mother although she also took some interest in this regard. Her school education ended in Class VIII and she got admission to the Medical College and Hospital for training in nursing but discontinued it. Ultimately, all roads led her to the world of music. Indubala began to receive training first from Gourisankar Misra, successor of the Benaras *gharana* of *saranga.* For some time, she was also tutored by Gauhar Jaan Bai. Indubala recalls, "She taught me under her personal care the manners related to high profile performance of music" (89). But Gauhar's tutorship had a short span and Indubala ultimately remained under the guidance of Kalisankar Misra. Simultaneously, with classical music, she concentrated on Bengali songs that opened for her a new scope of cutting gramophone discs under the Gramophone Company of India. Her first gramophone record came out in 1916 and later, she became the first Bengali woman singer to record Hindi songs that received appreciation from a wide cross section of people. Indubala started performing on radio from the very year (1927) that regular radio broadcasting started in Kolkata. She reached the zenith of her popularity and fame during the 30's and received a series of invitations from connoisseurs of music across India. She formed a concert party named Indubala Concert Party.[1]

She also established herself as a proficient actress in theatres and films, in which she bore her mother's legacy. Indubala recalls,

> My mother organised a theatre team composed solely of women. It was The Rambagan Female Kali Theatre… She did not form the theatre team for the sake of entertainment only or simply to satisfy her hobby. To help the poor and needy people, to raise funds for the cause of country… these were also its objectives. (92)

Indubala took part in twelve plays under the banner of this organisation that functioned for around three years. In course of time, she joined different theatre groups one after the other – Star Theatre, Kalika Theatre, Jupiter Cinema, Variety Palace, Minerva Theatre and Srirangam. In her reminiscences, Indubala drew attention to an important fact, "It was our 'Female Kali Theatre' that initiated the trend of organising theatre with women only – a trend that got currency in later years. So women of this day certainly owe their debt to my mother" (93). It needs to be mentioned that Indubala also acted in about forty films in Bengali, Hindi, Tamil and Telugu.

It is not the intention here to present Indubala as a renowned singer and actress but the background from which she rose to the height of fame may account for her empathy with neighbours. I would like to focus on her ability to uplift the conditions of her neighbours engaged in prostitution in the Rambagan area. No doubt, she received impetus for this arduous task from her mother and from her own struggles in life from childhood. In her sincere effort, Indubala left no stone unturned to make the best use of her talent. It may be pointed out that against the backdrop of the ani-*nautch* movement in different parts of India, the social mind in Bengal did not improve to the extent of offering status to women engaged in theatre and music even in the late nineteenth and early twentieth centuries. The drama movement in Bengal required actresses and prostitutes came forward to play roles in dramas with high hopes of earning a livelihood in the face of stiff opposition of society and news media. Girish Chandra Ghosh (1842-1912), a Bengali playwright and early contributor to drama music in Bengal, once lamented, "It is out of necessity that producers had brought women to the stage; but where would they get a woman of a good family

[1] Dr. Bandhan Sengupta. *Indubala*. Kolkata: Mousumi Prakasani, 1984.102-26. Print.

to be an actress in our society?" (Chanda, 258). In the very beginning of her autobiography, Nati Binodini, 'the prima donna of the Bengali Stage', confessed, "...socially I am a fallen woman, a hated prostitute" (1). At the same time, she had the guts to raise the question, "It is to be thought out who, for the first time, made life so despicable?" (Binodini, 53). An important turning point in Binodini's life occurred when Sri Ramakrishna, a pioneering personality in the religious revival movement in Bengal aiming to restore *sanatan Hindu dharma*, showered his blessings on her. It appears that Sri Ramakrishna's being a spectator of theatre contributed, to some extent, to the elevation of the social status of Bengali theatre in those days but could not do much to change the mindset of the conservative social elite. Brajendra Kumar Dey (1907-1976), one of the greatest composers of *yatra* scripts in the 20th century, correctly anticipated that womanhood would face extreme disgrace in the hands of commercially-motivated proprietors of *yatra* (Dey, 88).

> The expanding market of gramophone records contributed significantly to the popularisation of music in Bengal in the twenties (1923-25). But it could do very little to build a respectable image of performers like K Mallick, Bedana Dasi, Ascharyamoyee Dasi, Angurbala, Indubala and others in society. In his reminiscences, titled *Bratyajaner Ruddhasangeet,* Debabrata Biswas (1911-1980), a legendary *Rabindrsangeet* singer, observes: "Had we sung those sorts of songs (of Indubala and others) we used to face a severe scolding from our parents. They used to say 'What trash are you singing? Can't you sing Rabibabu's songs?'" (25). Debabrata's family was initiated into the Brahma cult that comprised people with an advanced social outlook in those days but senior members of his family were not in a position to appreciate the said performers in person or even their performances. The overall social situation was not congenial to facilitating a movement for upgrading the status of marginalised women in the 1950s, when Indubala came forward with her zeal to do something positive for them.

As far as her personal career as a singer and actress was concerned, Indubala received many awards, enjoyed prestige and love of her fans, earned a lot of money, could successfully overcome the ignominious identity she started with but could not remain satisfied with her personal achievements only. She waged a battle against the social evil associated with prostitution. Because of her empathy with women engaged in prostitution in and around the Sonagachhi area, Indubala associated herself with them and took active part in philanthropic initiatives. It was reported in the *Basumati* (9 Kartik 1329) that Rambagan Nari Samiti donated a considerable amount of money for helping flood-stricken people and also distributed rice among them. Another report in the same newspaper revealed that Rambagan Nari Samiti donated over seventeen hundred rupees to Acharya Prafulla Chandra Roy, the eminent scientist, through Indubala, who personally contributed four hundred and ten rupees.

What Indubala started at the age of twenty-three under the initiative and guidance of Rajbala continued in later years too. Sammilita Nari Samiti was an association constituted by socially ostracised women with the aim of ensuring betterment of their conditions. Its first conference was held on 22nd December 1956 at the junction of Premchand Baral Street and Banerjee Lane of central Kolkata. The *Jugantar* (26th December 1956) reported that Sri Sankar Prasad Mitra, the then Minister of Revenue and Judiciary, presided over the conference and Sri Jadabendra Panja, the then Minister of Cottage and Small Industries, acted as the chief guest. Sri Pratap Chandra Chandra was the main speaker. This conference appealed to the state government and the Government of India to 'adopt measures to protect the socially marginalised, rejected women engaged in prostitution from oppression and corruption, and to implement the urban programmes of the Central Social Welfare Board of the Government of India for the rehabilitation of socially marginalised women under the Second Five Year Plan'. The conference presented another proposal, that appealed to the state government and the Government of India, for inclusion into the Central Social Welfare Board two representatives of the Sammilita Nari Samiti who would cooperate to create an improved environment ensuring the mental, physical, financial and spiritual uplift of socially marginalised women inspired by the ideas of service, peace and emancipation'. It is important to note that in spite of social rejection, participants at the conference were inclined to share the social responsibility of donating money, clothes and rice to the Chief Minister's fund for flood-affected people. The conference also moved a resolution of condolence for the deaths of Dr Ambedkar and the wife of Dr Radhakrishnan.

The efforts of the marginalised women of Sonagachhi and adjacent areas did not end in 1956. The third conference of the Sammilita Nari Samiti was scheduled for 28th July 1958, when Indubala was about to turn 60. A handbill containing an appeal to attend the conference was circulated in Indubala's name. As reported by the *Jugantar* (30th July 1958), the two-day conference was attended by around 500 representatives from Howrah, Hooghly, Bardhaman, Asansol and different parts of Kolkata. The conference adopted a sixteen-point programme that emphasised the need for education of marginalised women's wards with dignity. Another resolution protested police oppression. It was further reported that the association had seven hundred members and it started collecting information about its members.

The effort to organise prostitutes for the sake of social and spiritual uplift might have engendered hopes for a better future. But the real story appeared far from encouraging. In a statement published in the *Jugantar* of 27th July, 1958, Indubala disowned the handbill circulated in her name and alleged that she had not even been informed about the details of the conference. Disgruntled, Indubala took the lead to form a counter social service organisation named Paschimbanga Nari Kalyan Samiti the aim of which was, as she described, 'to help the fallen sisters at this hard time'. Its office was located at 3, Jatindra Mohan Avenue of Kolkata. The ball rolled further. The executive committee of the conference, in an emergency meeting, reacted strongly to Indubala's statement in the newspaper and accused her of falsehood. The meeting went to the extreme of recommending her expulsion from the Sammilita Nari Samiti. Ultimately, she was expelled from the organisation with which she had been intimately associated since its inception.

What transpires from these incidents is that the political process that was ushered in after 1947 was gradually trying to bring society within its grip. The onset of economic planning, inspired by the ideal of a 'socialistic pattern of society', made it imperative for planners to make provisions for special homes and work houses for beggars, rescue homes for reception and rehabilitation of unmarried mothers, delinquent women and those rescued from houses of prostitution, and institutions for the benefit of the physically and mentally challenged. Contending political interests developed over these ambitious official programmes. It may be easily understood that Sammilita Nari Samiti became a victim of these inevitable and powerful political forces. Undaunted and spirited as she was, Indubala, jointly with Krishnavabini Debi, Saratkumari Dasi and nine others, issued a handbill in which they appealed to the conscience of sex workers:

> We do not think much about the idea that if we so intend, we can earn our bread by any means other than engaging in the flesh trade. You would certainly admit that we can even turn impossible into possible if we pay attention to it. Such a moment has appeared when all are trying to stand on their own feet. Why should we not? When we have no assigned place in the society of our country, we ourselves would have to build our own society. (Sengupta, 218)

Apart from extending this moral appeal, Indubala made contacts with the government many times so that concrete steps could be taken by the administration to rehabilitate women from red light zones. If we look back in time, we shall find that Indubala also rendered her service during the freedom movement of India. In March 1934, she actively participated in a women's assembly organised by the Indian National Congress and her speech drew the attention of the leaders. She had close ties with national leaders like Jatindramohan Sengupta and Bhupendranath Dutta and also provided shelter to the revolutionaries in Rambagan.

Indubala led a very colourful life – she was a singer, an actress on stage and in films, a successful organiser and a living inspiration to marginalised women of Kolkata. Mirabai led an ascetic life devoted to the adoration of Krishna and she was away from the so-called mainstream of the Rana's kingdom. She became 'people's Mira' at the end of her mortal life. But Indubala became a legend in her lifetime. At the award ceremony organised by His Master's Voice company on 6th April 1976, Tushar Kanti Ghosh, while introducing Indubala, wrote in a brochure published on the occasion, "Today it is an unbelievable chronicle that in an age when women's freedom was infringed on by hurdles and social sanction, when the participation of a female performer in the cultivation of cultural activities was hazardous, disrespectful, with what a turbulent and arduous endeavour Indubala established herself as an artist in those days of deception". In spite of all awards and good gestures, she could not erase the social stigma that she was a resident of an infamous red light zone. Nor was she willing to shed her identity as one of those who were compelled to carry on an ignominious profession. It was music that consolidated her position but she did not allow herself to be uprooted. She was very adamant when she confessed, "I am Rambagan's Indu. Here I have learnt music, established myself and got respect. Everybody knows that I am that Indu who belongs to Rambagan". Mira *bhajaniks* were both the cause and effect of Mira's myth. In fact, they recreated Mira after their own images. Indubala also remade herself as an agent of spiritual empowerment of women who fell from the grace of society.

Works Cited

Biswas, Debabrata. *Bratyajaner Ruddhasangeet*. Kolkata: Karuna Prakasani, 1394 (Bengali year). 25. Print.

Chanda, Pulak, ed., *Naribiswa*. Kolkata: Gangchil, 2008. 258. Print.

Dasi, Binodini. *Aar katha O Anyanya Rachana*. Kolkata: Subarnarekha, 1416 (Bengali year).1, 53. Print.

Devi, Indubala. "Ateet Diner Smriti". *Anandabazar Patrika*. Barshik Sankhya (1379): 88-9, 92-3. Print.

Engels, Frederich. *The Origin of the Family, Private Property and the State*. Moscow:Foreign Languages Publishing House, 1952. 122. Print.

Kumar Dey, Brojendra. *Jatrar Ekal Sekal.* Kolkata: Impression Syndicate, 1398 (Bengali year). 88. Print.

Mukta, Parita. *Upholding the Common Life: The Community of Mirabai.* Bombay, Calcutta and Madras: Oxford U P, 1994. 56, 66, 115, 130, 148-149, 168-169. Print.

Muhuri, Reba. *Thumri O Baiji.* Kolkata: Pratibhas, 2002. 13-5, 60-1. Print.

Sampath, Vikram. *'My Name is Gauhar Jaan!': The Life and Times of a Musician*. Rupa & Co., New Delhi, 2010. 88-90, 187-188. Print.

Sengupta, Bandhan. *Indubala*. Kolkata: Mousumi Prakasani 1984. 102-26, 218. Print.

CHAPTER TWELVE

SILENCE: A DANGEROUS SUPPLEMENT? RE-CONCEPTUALIZING CLASS/GENDER INTERACTION THROUGH THE LENS OF RABINDRANATH'S "PUNISHMENT"

ANIRBAN BHATTACHARJEE

While glancing over the piquant pages of *The Second Sex* (1949) for the first time with an insatiable gluttony, to be frank and candidly speaking, to know the un-knowable, I was, at times, taken aback by Simone de Beauvoir's occasional ascribing of ontological meanings to anatomical sexual differentiation. But a serious re-looking into the text reveals that anatomy alone does not have an inherent significance. As Beauvoir argues, natural facts actually gain significance only through their subjection to non-natural interpretation; following Maurice Merleau-Ponty's (1908-1961) logic of perception and affect, she puts: "man is not a natural species; he is a historical idea. Woman is not a completed reality, but rather a becoming, and it is in her becoming that she should be compared with men; that is to say, her *possibilities* should be defined"(Beauvoir 40). 'Becoming', in a sense, allows for the process of gender acquisition conceptualised as an 'interplay of individual choice and acculturation, which emphasises the constantly ongoing nature of the process itself' (Butler 1998 29). If gender is a way of 'existing one's body' and the *body* of the man/woman is understood as a *situation*, a field of cultural possibilities, both received and re-interpreted, the body then becomes a choice, a mode of enactment and re-enactment of received gender norms which surface as so many styles of the flesh. And my write-up would like to furnish, through one of Tagore's most anthologised stories, "Punishment", a woman's stylisation of her being that assumes certain corporeal significance, questioning the voluntaristic conception of *gender* as passively determined, constructed by a personified system of patriarchy or phallogocentric language which precedes and determines the subject itself.

In many ways, the most sustained challenge to class analysis as a central axis of critical social theory has come from feminists and the relative inattention to gender in the Marxist tradition is taken by many commentators as a de facto denigration of gender as a significant explanatory factor. However, a re-conceptualising of the explanatory project of class analysis invariably draws upon the central task to sort out for specific *explananda* the forms of interaction between class and gender in determining various outcomes. In this case, at least three forms of class/gender interaction can be considered particularly important: a) While the concepts of class and gender are analytically distinct, there are empirical situations in which gender relations themselves are a form of class relations. Frederick Engels (1820-1895) in his classic essay on the family and private property formulates the relationship between class and gender in early civilizations this way: "The first class antagonism which appears in history coincides with the development of the antagonism between man and woman in monogamian marriage, and the first class oppression with that of the female sex by the male …" (Engels 73) Indeed, control over the capacity of women to produce new labour power was one of the pivotal forms of property relations. b) Certain kinds of class positions may only exist by virtue of the fact that specific forms of gender relations are present. Critics have analysed that in our patriarchal households, gender relations play a crucial role in making possible maid and childcare services. (Olin Wright 119) Interestingly, the notion of *work* which is basically premised upon *commodification* and selling of *labor-power* in a capitalist mode of production is made possible because of the existence of a family where women would take care of the household work. When in a capitalist class-machine, the labor-time of the (male) worker is purchased at a cheap rate through a discursive forgetting of the labor-power that produces the commodity; keeping the hegemonic structure intact, the task of a woman in a hetero-patriarchal dwelling always hinges upon some 'unproductive' labor economy. The institution of family founded upon certain conceptualisations of women in managing the household is, therefore, crucial to the maintenance of class hierarchy. c) Again, the distribution of power and resources within gender relations affect the likelihood of men and women occupying certain kinds of class relations. Individuals are linked to class structures through a variety of relations other than their direct location in the social relations of productions. However, in this mechanism of multivariate interactive effects always mediated by a certain kind of differential economy, the 'woman question' remains a problematic that must re-surface with a proliferating polity of gender, body and class.

Manu Smriti, a controversial legal literature, because of its theory of *Varna-dharma* for social order and protection of creation based on *Varnāshram* as a rule of law, propounds eternal laws for a husband and his wife, in which "women must be kept in dependence by the males of their families and, if they attach themselves to sensual enjoyments, they must be kept under one's control" (9.2). It further asserts, "a woman should be protected by her father in childhood, by her husband in youth, and by her sons in old age; a woman is never fit for independence" – *na stree svātantram arhati* (9.3) (*Manusamhita* 698). Interestingly, Sibaji Bandyopadhyay in his article, "Producing and Re-producing the New Women: A Note on the Prefix 'Re'", shows how the 'family', which has been organised as a 'sphere of merger', a site of amity and fellow-feeling in a competitive and individualistic bourgeois model society, gets distorted beyond recognition in the colonial hell: The colonial state, as a part of that subterranean 'base' upon which the 'superstructure' of capitalist economy was built, could never allow the establishment of an 'affirmative relation' between 'family', 'civil society' and 'state' in the colonial societies (23). It is the prison-house of belonging in which a woman's seeking of transcendence from the repetitive pattern of life fringes upon the border of the 'mis-known', the unanticipatable *arrivant*.[1]

We will now probe into the crafty mechanism of social hegemony in a typical agrarian Bengali society to locate a woman's voice that plays upon her own *being*, almost in a self-deluding fashion, questioning and, at times, parodying the dominant social system. To somehow subvert and undermine the sanctity of an official discourse, in my small-scale maneuver, I am taking up one of Rabindranath's most anthologised stories, "Punishment".

The action in "Punishment" unfolds thus: Two peasant brothers, Dukhiram Rui and Chidam Rui, and their wives share living accommodations as subtenants of the village know-all about legal matters,

---

[1] For Derrida, the *arrivant* signifies the 'perhaps' of the 'what arrives'. As he puts it: "What is going to come, *perhaps*, is not only this or that; it is at last the thought of the perhaps itself. The *arrivant* will arrive perhaps, for one must never be sure when it comes to arrive; but the *arrivant* could also be perhaps itself, the unheard of, totally new experience of the perhaps" (Derrida 2000). In Derrida's radicalised notion the *arrivant* is affirmed as an "absolute alterity", meaning that the absolute must be permanently absent. The *arrivant* does not come to a place named or determined in advance, but rather "affects the very experience of the threshold".

Ramlochan Chakrabarti. Indeed, the layering of tenancy, fragmentation of landholdings and landlessness were endemic in rural life in colonial India, and in Bengal in particular, it became progressively worse from the Permanent Settlement of 1793, as detailed in a report[2] on the progress of land reforms in Bengal following Independence. The two brothers return from a long day working on the zamindar's office-building, having been forcibly withdrawn from cutting paddy by the bailiff (and Ranajit Guha describes how forced labour was a feature of rural life in colonial India[3]) to find their wives idling at home. Dukhiram, the elder brother, demands food from his wife, Radha, and when she protests mockingly that he hasn't provided any food for her to cook, he loses his temper, lashes out and kills her with his sickle. Ramlochan arrives to collect his overdue rent and Chhidam, the younger brother, to cover up his brother's deed, unwittingly identifies his beautiful wife, Chandara, as the perpetrator. Prior to her imprisonment pending trial, Chhidam instructs Chandara to admit to the murder but to say it was in self-defence. Out of stubborn pride, she simply tells the police and later the magistrate, 'Yes, I killed her', and denies there was any brawl or attack to provoke or excuse the crime and she is condemned to death. In fact, she accepts the injustice of a punishment for a crime she did not commit in order to punish Chhidam for forcing her into an intolerable position and does not give him the satisfaction of saving her. She wields the ultimate weapon of sedition by refusing to see him before her execution, contemptuously exclaiming, '*maran!*'- a term which, in literal rendering, means 'death'. But it actually underlies a host of complex and untranslatable implications: Anger, exasperation, hatred and a certain kind of suppressed eroticism.

Again, if we step into the contending factors underlying an otherwise embedded discourse based on some 'naturalistic' assumptions about the capacities and desirable roles of men and women, which may stand closer to the *Manu*ist ideals of a breadwinning husband supporting a financially dependent wife, it will possibly create some 'dissenting spaces' making the semiotic order wholly unstable. After Chandara's marriage to Chhidam

---

[2] Karunamoy Mukherji has discussed it in his article, "Progress of Land Reforms in West Bengal" published in *Rabindrabharati Journal* [Vol.I, July 1968, pp.58-76].

[3] While writing into the history the hegemonic consciousness and political agenda of Indian peasants and urban factory workers, whose revolutionary anti-colonial and anti-bourgeois political voices were suppressed in Indian historiography, first by the colonial writers and then by the Indian elite nationalists, Ranajit Guha describes in detail the exploitation of *beggar* or forced labour by every dominant group, including Landlords. (Guha 1997 26-30].

at an early age, her father, at death-bed, expressed his deep-felt content, saying: *Jaha houk, amar meyetir ekti sodgoti koriya gelam!* ['Whatever happens, I've provided for my daughter'].

Now we will undertake a close reading of each and every move of the woman, *a question*, to unfold if she remains 'bound' to 'the authoritative wor(l)d' (Bakhtin 342) in its totality or somehow exercises her potential to react to the many assumptions of the male culture. There is no denying that women, as Wollstonecraft wrote in her 1972, *A Vindication*, "because they have not been led to consider the knowledge of their duty as the one thing necessary to know, or to live in the present moment by the discharge of it, they are very anxious to peep into futurity, to learn what they have to expect to render life interesting, and to break the vacuum of ignorance" (205-6). The question is, if this woman has attained the exacting standards of 'refinement' to loiter in the land of 'no-land' or if she is only the 'natural nurturer' (Bandyopadhyay 1994 22) supplementing the authorial discourse. Indeed, Chandara is one of the few women in the Tagorian 'writing back' oeuvre who, in spite of her low-class, humble origin, raises her voice against the oppressive patriarchal order, adopting certain modes of inversion and subversion.

But is Chandara's voice audible? Is the question of 'class' always already imbricated within the gendered role of the persona? With a backward look into the familial bond before this unprecedented happening, we see Chandara suspecting her husband of infidelity and beginning to flirt at the watering hole. He then threatens her with bodily harm and locks her in the house. She escapes to a relative's house but is persuaded to return only after Chhidam 'has to surrender to her'. Tagore relates, "It was as hard to restrain his wife as to hold a handful of mercury" (110-20), but contradicts this description with the assertion that Chhidam "did not have to use force any more" to control his wife. Chandara has achieved a sort of power by submission, itself a paradox. Where the balance of power lies in this relationship is almost uncertain. But while we niggle over the balance of authority between personalities in the Rui household, we should remember that "the entire culture supports masculine authority in all areas of life – and outside of the home – permits female none at all" (Millet 35). Patriarchy is the governing ideology in which the male figure is both the begetter and owner in a system in which kinship is property. The continual surveillance over Chandara, "a tiny defenseless flighty laughter-loving village wife" attests to the fact that in patriarchy "a large quantity of guilt attached to sexuality is overwhelmingly placed upon the female who is, culturally speaking, held to be the culpable or more culpable party in nearly any sexual liaison, whatever the extenuating circumstances" (Millet

54). Indeed, a tendency towards the reification of the female makes her more often a sexual object than a person. However, it is interesting to note that unlike men, who assume a closure of their identity that is already a given in a patriarchal social system, for women, there is no home, no pre-given identity and no centrality of subjecthood who can know and who can be known. She is always in a process of experiencing and *gathering her own being* trapped within a phenomenal chain of events. Her knowledge of the world, too, is grounded in a shifting being that 'she' is – the immediacy of her experiences has a meaning in the world, but that meaning doesn't belong to her fully.[4]

Chandara's compulsive attachment to her coy consort that carries an air of her belonging to an everyday world gets, therefore, ruptured by this untoward happening, revealing her sheer sense of non-belonging brewing from the beginning of the marital bond. It seems that her troubled phenomenal existence in a *no exit* situation has always been haunted by an uncanny *arrivant* of death symbolic of exit. The question bubbles forth: Where lies the transcendence of the woman, her escape from this *mayik* structure of belonging/unbelonging?

Max Weber holds the view that property or 'lack of property' is the basic category of all class situations (182). But in the entire structure of agrarian Bengali society, it is the 'status segregation' that accelerates the formation of 'caste'. The caste-structure transforms the horizontal and unconnected coexistences of ethnically segregated groups into a vertical social system of super- and sub-ordination. In this status segregation, women are considered in the male conscience as 'the burden of economic dependency' (Millet 37). As Kate Millet puts it, it is through this system that "a most ingenious form of 'interior colonisation' has been achieved. It is one which tends, moreover, to be sturdier than any form of segregation and more rigorous than class stratification, more uniform, certainly more enduring" (54). As far as patriarchal religion, popular attitude and the psycho-social structure of the family are concerned, women, though granted legal citizenship, tend to be ruled through family alone and have no formal relation to the state. But the interesting point is that one should *not* consider himself/herself a victim of the hegemonic operation in order to de-center the constructed edifice built upon some unsure ground.

[4] Samrat Sengupta interestingly poked out this problematic in a paper titled, *Is it Possible for Women to 'Cease Upon the Midnight with No Pain': Understanding the 'Masculinity of Suicide' as Resistance*, presented in 'Agency and Resistance: Feminist Approaches' conference held at Women's Studies, Jadavpur University on 13-14 March, 2012.

Chandara adopts certain guerilla-like tactics of *silence* and *repetitive negation* to undermine the sanctity of legal and official discourse.

Law exists, as Weber has analysed, "when there is a probability that an order will be upheld by a specific staff of men who will use physical or psychical compulsion with the intention of obtaining conformity with the order, or of inflicting sanctions for infringement of it" (180). Weber argues that we understand, by 'power', the chance of a man or a number of men to realise their own will in a communal action even against the resistance of others who are participating in the action. The chain of events after the murder actually exposes the complexity of the legal discourse, labouring away on new meanings of the question of 'telling the truth'. The agreement between husband and wife is that Chhidam will save Chandara from execution while she acquiesces to his lie. Chhidam expects Chandara to relate that her sister-in-law attacked her and was killed (by Chandara) in self-defense. But after being taken into custody by the police, Chandara tells them the attack was unprovoked and puts her own life at risk by defying her husband. At the fag end, Dukhiram's protestations of his sister-in-law's 'innocence' serve only to convince the court of her culpability, reversing the notion that 'the truth shall set you free'. Chandara's unassailable stratagem of *repeated negation* actually complicates the aporetic structure of law and justice, in which 'the justice of law, justice as law is not justice'.[5] Justice remains always *to-come,* it remains by coming; there is no horizon of expectation.

The narrative, a polyphonic battle royal, unfolds a turmoil of issues: Dehumanising poverty, forced labour and fragmentation of land, *peasants incapable of enforcing their class interests in their own name*[6], the family

---

[5] Jacques Derrida quotes Montaigne in distinguishing laws from justice: In *Acts of Religion*, he analyses why/how law is not justice. "Law is the element of calculation, and it is just that there be law, but justice is incalculable, it demands that one calculate with the incalculable; and aporetic experiences are the experiences, as improbable as they are necessary, of justice, that is to say of moments in which the decision between just and unjust is insured by a rule". (Derrida 244).

[6] In *The Eighteenth Brumaire of Louis Bonaparte* (1852), Karl Marx observes: "In so far as millions of families live under economic conditions of existence that separate their mode of life, their interests, and their culture from those of the other classes, and put them in hostile opposition to the latter, they form a class. In so far as there is merely a local interconnection among these small-holding peasants, and the identity of their interests begets no community, no national bond, and no political organisation among them, they do not form a class. They are consequently incapable of enforcing their class interests in their own name, whether through a

and legal structure, murder, perjury, love, judgment, the individual and the universal, untranslatable words and cultural differences, most of which are inextricably enjoined with the question of Chandara's *becoming*. It projects a purposive and appropriative set of *acts* to assume a certain corporeal style and significance. The linguistic route, in Chandara's case, actually moves along this line: Clamours and quarrelling, repulsion to talk, repeated rant of the word 'No' culminating in *silence*. Indeed, the linguistic stylisation of her *being* gets a new turn as she starts pondering over the unforeseeable haunt of *thanatos*.[7] But it is in this moment of *stasis* that she is actually fashioning her own being; silence becomes the embodiment of her denial, her resentment that questions and parodies her own *situatedness* within a classed/gendered hegemonic discourse. When the police start investigation asking her: 'Why did you kill her?' 'Was there any quarrel?' 'Did she try to strike you first?' 'Had she oppressed you in any way?' and more blatantly, 'Do you know the punishment for the crime you confess to?' Chandara condemns herself to essential muteness, affecting the rustles of phalocentric language.

Here, one needs go deeper into those lousy and intricate patterns of self-making a woman adopts to break any sure and secure signifying chain. Chandara's silence is not a simple absence of speech; it is the inverse of language where the interlocutor flings off a sign, but declines any interpretation; it is the silence that terrifies. Certainly, these *silences*, in the *iterability* of their structure, have been *performatively* undermining the structures of patriarchal discourses.[8] Chandara's world of silence is 'an

---

parliament or through a convention, they cannot represent themselves, they must be represented".

[7] It was originally proposed by Sigmund Freud in 1920 in his fascinating *Beyond the Pleasure Principle* where in his first published reference to the term, i.e., *Thanatos*, he wrote of the "opposition between the ego or death instincts and the sexual or life instincts". With this essay, Freud went "beyond" the simple pleasure principle, developing his theory of drives with the addition of the death drive(s) – 'the hypothesis of a *death instinct*, the task of which is to lead organic life back into the inanimate state'. Freud identifies these two drives that both coincide and conflict within the individual and among individuals.

[8] Anirban Das addressed this question of 'silence' in a paper titled, 'Feminist Politics, Theory and the Question of the Outside: Reflections" presented in the "Feminist Inscriptions in Social Theory" conference held at Centre for Studies in Social Sciences, Calcutta on 22-24 February, 2013.

Drawing upon the question of patriarchal/feminist sexualisation, Das says, "The silence I am speaking of is not at the cost of speech. It is the inalienable constituent of speech. Inalienable, yet interrupting. Without these interruptions, speech cannot work. Speech gets fixed. Silence, here, is something that makes speech operable

an-archic, maddening one, at the limit of non-sense in which its presence in consciousness lies in its expectation for a word that does not come' (Levinas 82-90). The incorrigible rant of the woman wrapped up in the depth of silence is like laughter, perfidiously held back, that seeks to destroy language. No ontology can think of the operation of this discursive silence that haunts leaving its trace without ever being present or absent, and thereby transforms the terrain of reason. The *being* of the woman is, in a way, caught up in the movement of supplementarity entailing an originary form of *auto-affection* – a primary narcissism – standing outside the relations of power, having some subversive potential. When life seems to be a closure within which one is hermeneutically chained, Chandara's urge for an auto-effacement jars the *repressive* patriarchal meaning-making apparatuses. However shaky, almost ignored, but a purposeful, disguised kind of *performativity*[9] of a 'gendered' being, somewhat darkened by her humble 'class' position, throws a radical challenge by unswitching herself from the existing politic of patriarchy. Yet, the crucial problem remains: Chandara's rejection of a certain pattern of life invariably foregrounds the very constructedness and contingency of that pattern. The question is: Does the strategy for an *exit* lie in a woman's *everyday* denial of a repetitive economy, having the foreknowledge of the constituted immediacy of her elemental finitude, which pretends to belong to the law of the *logos*, without ever belonging, for: *To utter the longest of sentences, one has no need of words*![10]

---

through interruptions. This humiliation is not of the woman who suffered. It is that of whom who witness the suffering. The responsibility and care of the feminist witness".

[9] Butler's theory of 'performativity' seeks to explain how the subversion of power emerges within a dialectical relation between constraint and agency. She underscores gender's constructed nature in order to fight for the rights of oppressed identities, those identities that do not conform to the artificial – though strictly enforced – rules that govern normative heterosexuality (Butler 1993, 2004). As concerns, my notion of *alternative performativity* that I have brought up in this piece, basically attempts to access the *silence* /ing of the *gendered* being, her mode(s) of self-fashioning vis-à-vis death/ exit/signing out – a phenomenological understanding of her constant combustion, so to say – that can shake aside the repressive *shastrik* meaning-making apparatuses.

[10] Jacques Derrida quoting Charles Baudelaire in his The *Animal that Therefore I Am*, the long-awaited translation of the complete text of Derrida's ten-hour address to the 1997 Cérisy conference entitled "The Autobiographical Animal".

Works Cited

Bakhtin, M.M. *The Dialogic Imagination*. Ed. Michael Holquist. Tr. C. Emerson & M. Holoquist. Austin and London, 1981. Print.

Bandyopadhyay, Manubendu, tr. & ed. *Manusamhita*. Sanskrit Pustak Bhandar, Kolkata, 2010. Print.

Bandyopadhyay, Sibaji. "Producing and Re-Producing the New Women: A Note on the Prefix 'Re'". *Social Scientist*. 22. ½ (Jan-Feb.1994):19-39. Print.

Butler, Judith. "The Lesbian Phallus and the Morphological Imaginary". *Bodies that Matter: On the Discursive Limits of "Sex"*. New York & London: Routledge, 1993. Print.

—. "Sex and Gender in Simone de Beauvoir's *The Second Sex*'" in *Simone de Beauvoir: A Critical Reader*. London& New York: Routledge, 1998. Print.

—. *Undoing Gender*. New York & London: Routledge, 2004. Print.

Chatterjee, Partha. "Caste and Subaltern Consciousness".*Subaltern Studies VI*. Ed. Ranajit Guha. Delhi: O U P, 1989. 169-209. Print.

Das, Anirban. 2013. "Feminist Politics, Theory and the Question of the Outside: Reflections" paper presented in the *Feminist Inscriptions in Social Theory"* conference held at Centre for Studies in Social Sciences, Kolkata on 22-24 February, 2013.

Beauvoir, Simone de. *The Second Sex* (1949). Tr. H.M. Parshley. London: Pan Books, 1972. Print.

Derrida, Jacques. *Aporias: Dying – Awaiting (One Another at) the 'Limits of Truth'*, tr., Thomas Dutoit, Stanford: Stanford U P, 1993. Print.

—. "Passages – from Traumatism to Promise". Tr. Peggy Kamuf. *Points...Interviews, 1974-94*. Ed. Elisabeth Weber. Stanford: Stanford U P, 1995. Print.

—. 2000. *Of Hospitality*, tr., Rachel Bowlby, Stanford: Stanford U P.

—. "The Animal that Therefore I Am". *Critical Enquiry*. 28.2 (Winter 2002): 369-418. Print.

—. *Acts of Religion*. Ed. Gil Anidjar. New York & London: Routledge, 2002. Print.

Doniger, Wendy & Brian K. Smith, trs., *The Laws of Manu*, New Delhi: Penguin Books, 1991. Print.

Engels, Frederick (1884). *The Origin of Family, Private Property and the State*, Resistance Australia: Marxist Library, 2004. Print.

Freud, Sigmund (1920). 'Chapter VI', *Beyond the Pleasure Principle*, tr., James Strachey, London: Vintage, 2001. 44-61. Print.

Guha, Ranajit. *Dominance without Hegemony: History and Power in Colonial India*, Cambridge Massachusetts: Harvard U P, 1997. Print.

Lacan, Jacques. "The Signification of the Phallus". *Ecrits*. Tr. Bruce Fink. New York and London: W.W. Norton and Pantheon Books, 2006. 575-584. Print.

Levinas, Emanuel. *Totality and Infinity*. Tr. Alphonso Lingis. London: Martinus Nijoff Publishers, 1979. Print.

Marx, Karl (1852). "The Eighteenth Brumaire of Louis Bonaparte". *Karl Marx: Selected Writings*. Ed. David McLellan. OUP, 977. Print.

Millet, Kate (1969). *Sexual Politics*. London: Vergo Press, 1977. Print.

Montaigne, Michel de. *The Essays of Montaigne*. Tr. John Florio, New York: Modern Library, 1993. Print.

Mukerji, Karunamoy. "Progress of Land Reforms in West Bengal", *Rabindrabharati Journal*, Vol.I (July 1968): 58-76. Print.

Olin Wright, Erik. *Class Counts*: *Comparative Studies in Class Analysis*. Cambridge U P, 1997. Print.

Sengupta, Samrat. "Is it Possible for Women to 'Cease Upon the Midnight with No Pain': Understanding the 'Masculinity of Suicide' as Resistance", paper presented in *Agency and Resistance: Feminist Approaches* conference held at Women's Studies, Jadavpur University on 13-14 March, 2012.

Tagore, Rabindranath. *Shasti* ("Punishment"). Tr. Supriya Chaudhuri. *Selected Short Stories*. Ed. Sukanta Chaudhuri. New Delhi: OUP, 2000. 110-20. Print.

Weber, Max. "Part II: Power-Sec VII: Class, Status, Party". *From Max Weber: Essays in Sociology*. Tr. H.H. Gerth & C. Wright. Oxford: O U P, 1972. Print.

Wollstonecraft, Mary.1972. *A Vindication of the Rights of Women*, London & Vermont: Everyman, 1995. Print.

CHAPTER THIRTEEN

PROCESSES OF GENDERING AND SCHOOL AS A MODERN FOUCAULDIAN INSTITUTION: SOME PERSONAL REFLECTIONS

SREENANTI BANERJEE

> "The political question […] is not error, illusion, alienated consciousness or ideology; it is 'truth' itself – Michel Foucault.

Description is not an innocent act. It is precisely in and through the act of description that we *constitute* a discursive 'outside', the 'object' of my discourse. This paper too forms its own discursive outside, the school, and thus certainly has chances of producing it as a somewhat reified and 'finished' category or a product. However, here I must hasten to add that this description certainly does not exhaust all the *possible* ways in which I might look at the schools I went to, and there are other ways of 'describing' (constructing) as well, which, once again, would form their own 'realities', some of which might overlap with that of mine, some might just exceed, undo or subvert my act of *constituting* the object of analysis. Thus, here the category 'school' should certainly be construed as a *constructed* 'text' and not a determined fixed object, where any effort to construct will always and already bear the traces of other kinds of constructions too.

The two high schools I went to were Carmel Convent (Kolkata) and Patha Bhavan (Kolkata), the latter being where I pursued my higher secondary education. The two schools were engaged in their own 'truth' making practices and *rituals*, their own ways of 'humanisation' and 'subjectivation'.

However, what I intend to do here, is not to engage in a process of 'revealing' or 'unpacking' the 'real' nature of the schools hidden behind its apparent humanising veneer. These institutions '*recruited* me as a *subject*'[1]. Hence, they could have been anything but certainly not a 'lie', a 'repressive mechanism' or a 'negative ideology' which 'corroded' my *true authentic sovereign 'untainted' pre-school Self.* My Self, as I know it today, is a 'result' of the 'liberties' I was ritualised with for the first 16 years of my life. It could have been anything, but certainly not a mere act of falsification. Here, when I use the word 'Self' in terms of a 'subject', I certainly use it in the Foucauldian sense of the term, something that would mean both subjection (in the sense of 'being subjected to') and subjectification (in the sense of being an 'agent'), referring to the two meanings of the word, *subjectum* or a radical initiator of things and *subjectus* or annihilation/destitution of the Self. Interestingly, there are several such words associated with my experience of schooling that are punctuated by this *double bind.* The aim of this paper is to explore some of these conundrums and to trace a genealogy of how certain supposedly opposing, incommensurate essences could actually come together.

I. Disciplining the Body to 'Freedom': Authority as Spectral

> "Within thy hallowed portals, Carmel dear, we learn thy *truth* that casts out fear.God is our all, he's always near, thus inspired with *faith* and *wisdom* clear. Raise then on high, the Carmel banner bright, to one and all it shall proclaim. We stand for truth, for *equal rights*, in prayerful trust and *in God's name*" (emphasis mine)."

The above-mentioned lines from my school anthem, "we stand for *truth*, for *equal rights* [...] in 'God's name', allude to a significant cohabitation. Interestingly, this urge to adhere to the notion of 'truth' and 'equal rights' or in other words to be 'enlightened', self-judging and self-reliant certainly refers to the Kantian idea of *Aufklarung* or Enlightenment, where an individual uses his own capacities of 'pure' reasoning, and not that of the pastor or any other authority, and 'owns up' to responsibility for decisions taken and judgements given by him. Enlightenment here is a *result* of the age old confines of religious theocracy, it is a 'way out', and hence, certainly should not be construed as an 'originary' intending act of the self-same rational subject. It is an act of 'becoming', where one 'becomes' a subject using the capacity of the *decision of his own will*.

Here, this subjectification of the student in the public sphere (where 'wisdom' ought to be exercised) was always refering to a kind of sovereign power, a force resonating with Kant's famous words, "Argue as

[1] A phrase borrowed from Louis Althusser's *Ideological State Apparatus* (1993).

much as you can, but *obey*", a force which ensured that we got regulated by an authority. We argue, exercise our autonomy and freedom of reasoning but we do this for an invisible anonymous authority that the anthem deems 'God'. While in the *public* sphere, I was supposed to reason, that reasoning was legitimate only to the extent that in my *private* sphere I get ritualised to the meaning of 'authority'.

While as a child I was taught to realise that this authoritarian figure was supposed to be God, as I grew up, this authority became anonymous, it spread its wings in the form of a ubiquitous capillary, and its 'origin' could not be traced. But, the *spectre* of this 'originary' authority remained for life. I came to realise that I could only gain the right to reason rationally provided I met the criteria of being a 'responsible' mature citizen-subject certified by that spectral authoritarian figure, the 'expert', or the sovereign, who himself/herself is *subject* to the same vocabulary of disciplined rationalisation, just like me (a realisation which occurred much later). I gathered that the ethics of ensuring 'equal rights' (something that the words of the anthem echoed and a prescriptive message that was passed on to us every day in school) would only work provided one was 'fit and deserving enough' to gain those rights. Thus, the school siezed the whole 'life' of the student, in its totality, where certain aspects, traits or persons were considered inimical to the notion of 'good' and 'rational' *life* itself. Thus, the processes and *tactics* with which the school aimed to *make us live* were the same discursive tactics which also *let other 'rogued' aspects of ours die* (*Society Must Be Defended* 240).

II. Realising the Universal *in and through* the Concrete

One of the significant aspects of Foucault's philosophy is the genealogical analysis of how transcendental universal (analytic) categories, like say that of a 'good intelligent student' or a 'respectable woman' (the 'unbound serialities' of Benedict Anderson in some sense) are constituted in and through the act of disciplining and *governing* (resonating with the notion of 'bound seriality'), where the universal is passed through the grid of the 'descriptive' *concreteness* of bodily practices, or in other words, the very act of *detailed* descriptions of what one 'ought to do' also constructed the *telos* of the discourse. And, most interestingly, this was also directly related to the manner in which abstract (mathematical) concepts were realised (for instance, the abstract concept of say 'Number 12' was explicated through the notion of 12 tables (Chatterjee 130).

Something similar happened in the school as well where the 'norm' was realised in and through disciplinary practices. For instance, a 'welfare week' was arranged by the authorities every month when students donated money or things (food, stationary items, clothes, etc) for the '*helpless* poor' (a practice which vigorously *depoliticised* the notion of 'need' in terms of inculcating a sense of the 'victimised' poor being an essential ontological category, and not a result of structures, often compared with a physically challenged person, where both were lumped together under the larger rubric of the 'needy', something that certainly created a discursive culture of *benevolent philanthropy*). Thus, we were taught to be self-sacrificing, rational, 'giving' agents. Thus, this culture of individualism (I am 'responsible' for whatever happens to me) informed the processes of gendering to a great extent. Structural oppression was turned into individual 'misfortune'. The moral science classes (which were also known as 'value education') were replete with such instructions of how, only by being 'well-behaved' in our demeanor and ways in the public sphere, we could earn a good reputation as young women.

Another significant aspect was the manner in which students were vigorously fined for every vernacular word that they articulated. Nobody was exempted from this rule and the authorities were extremely stringent about it. This would later get translated into vigorous neo-colonial presuppositions and resultant vilification of all women (and people in general) who would not be able to meet the demands of such quotidian habits in terms of the langauge they would be able to use in their daily chores.

Moreover, sovereign power of the authorities became visible only during situations of 'exception' (although what is at stake here is certainly not how tolerant the teacher was to difference, but rather the fact that what got defined and justified as 'exception' was certainly informed by normalising power, the demands of reaching upto the *desirable signified* of 'life' itself). In all other instances, the teacher could not exercise her personal whim as s/he was subjected to the same rules of discourse as the students themselves, *transparently* coded in the school constitution (although since s/he was designated as an 'expert', her subjective opinions were certainly translated into objective public 'truths', open to public scrutiny). Here, the individuals in their 'positions of responsibility' felt themselves to be free precisely because they were classified. This certainly resonates with the parable of the Law of Kafka, where between the guardian of the Law and the man from the country, there is no essential difference, since they are placed in antithetical but symmetrical positions, where nobody knows who the 'real guardian of law' is. Thus, such a functional aspect of power inculcated a culture of vigorous 'individualisation of the Self', in order to approximate the sought-after *telos* or the 'norm' of a 'good allrounder' (something that also aimed at bringing the future within the realm of the *calculable* present). These abstract mechanisms of 'taming' the Self impacted my notions about myself to a considerable extent.

A significant instance would be the school library where space was organised in such a manner that it inculcated a culture of a supposedly *democratic* cultivation of silence. Not only was the democractic art of reading silently cultivated but along with that, seating arrangements were also organised in such a way that readers had to sit separately from one another and not face to face. All of this facilitated privacy as well as a sense of self-surveillance and personal responsibility or an ethic of the 'Care of the Self' (in terms of borrowing books, returning them on time, paying fines, etc), certainly resonating with the Foucauldian notion of the *Panoptiocon*, where a sense of democratic movement was cultivated only to the extent of vehement and silent atomisation of individuals.

Thus, here we were 'subjected' to our own 'visibility', and thus had to be 'known' on all fronts. And, since we were aware of the fact that we are being observed, our bodies were rendered docile and compliant in the process where the act of enlightening us was not violating, but rather acting as a 'multiplier' – a multiplier of roles, personalities and hence, effectivity. It enhanced our 'productivity', 'usefullness', 'intelligibility' or 'transformability' (thereby vigorously inducing the notions of the *homoeconomicus* in us), *distributed* us more effectively (by, say, dividing us into groups during workshops) and thus, facilitated the processes 'identification', in order to further the process of supervision (Foucault *Lecture 8* 170).

The disciplinary tactics also aimed to make the most out of 'time', in and through the timetable or the regularised, periodical ringing of bells to mark the end of one class and the beginning of another. Thus, time was accumulated, divided into an organised plan and made to function in a *series.* It penetrated the body and aided it to engage in a rhythm. Hence, each of us got ranked based on the efficiency with which we could engage in a repetition of several tasks, something that also aided the process of *intervention* in order to correct (or 'educate' the student) the process of repetiton of the allocated tasks. Thus, we *embodied* the rule and the norms, a 'norm' which was instituted to measure and rank students, something that homogenised us by rendering pure differences as measurable 'deviations', and at the same time individualised us by assigning an 'identity' (Foucault *Discipline and Punish*).

Along with this, we were also instructed to wear 'knee-length' skirts, a rule which was almost always flouted, and students adopted various mechanisms to escape the dictums, one of which was the *strategy* of wearing the skirt as it wass so that it looked long and 'decent' during the assembly when the 'uniform-check' was on and folding it near the waist in order to shorten the length at other times. Thus, one thing that ran through this entire scripture of regimented humanisation is the notion of 'resistance'. This notion of resistance is remarkably explained by the eminent postcolonial critic Homi Bhabha (pp. 246). According to him,

> "Resistance is not necessarily an oppositional act of political intention, nor is it the simple negation or exclusion of the 'content' of another culture, as difference once perceived…(but) the effect of an ambivalence produced within the rules of recognition of dominating discourses as they articulate the signs of cultural difference."

Thus here, there was resistance, precisely because power and the processes of disciplining were ubiquitous, productive and generative.

III. Fixing the Fictive: Interrogating the 'Origin'

If Carmel was this supposedly individualising, atomising, Westernised space of materialism which apparently put forth the notion of an *adequately* feminised, globalised *presentable* citizen-subject, considered *smart enough* for the neo-liberal public-sphere, Patha Bhavan came with a promise, the promise of providing its students with an 'ethical, spiritual, non-Westernised' yet 'elite' space, punctuated by 'our traditional *civilizational*' legacy.[2] It was told to me that learning here was not a mere assimilation of knowledge, but about cultivating a *sensitive* consciousness about the surrounding environment, about 'nature'. Thus, it offered subjects like Fine Arts and Music; certain classes were held outside the confined premises of the classroom, most of the time in small groups where students would sit in semi-circles, more in the format of a seminar. Thus, it aimed to create an atmosphere of supposed 'democracy', free thinking and non-hierarchy where there was an apparent movement from *formal* distribution in space to a 'cooperative' *participatory* learning ambience, from *hierarchical* observation to 'differentiated dynamic'

[2] Dipesh Chakrabarty. *From civilization to globalisation: the 'West' as a shifting signifier in Indian modernity*. Inter-Asia Cultural Studies, Volume 13, Number 1, 1 March 2012, pp. 138-152(15).

learning and from *normalising* judgment to 'authentic' evaluative processes.

It is interesting to note here that the lecture mode, which was adopted in Carmel, informed by the notion of an apparently non-reciprocal power relationship, is less dubious than the seminar-mode of Patha Bhavan. A lecture is hesitant about its truth claims as it exposes itself to interrogation by rendering the power relations more visible; whereas the apparent non-hierarchical reciprocity of the seminar may have the veneer of participation, but in actuality conceals power relations to such an extent that although students assume that the situation is supposedly equitable, they, in fact, are passive consumers of the ambience created by the teacher, are more under vigilance are hence, increasingly self-surveiling. Such situations also demand more performance due to increasing amount of peer pressure.

Thus, this school too, had its own ways of *intensifying* subjectification and humanisation. While it disseminated a vocabulary of equity, democracy and what we know as 'high culture', it was also heavily invested in the making of what we know as the *bhadramahila*, the 'gentlewoman' – elite, enlightened, yet gentle and feminine. Here, I would like to bring in a radical Foucauldian *decentering* of the notion of the 'subject' (and an associated Foucauldian critique of Humanism as well) that the school tried to bring forth by proposing that this (civilized and cultured) 'subject' who was supposed to 'hold on' to the legacy of 'our' civilization was always and already *marked*, and this was certainly a problem of 'Western Humanism' itself, a Humanism which was heavily invested in the idea of a *full presen*ce or an *originary sameness* of *identity* which thereby was thought to get *tainted* by the onslaught of Westernisation and globalisation, something that was thought to remove our legacy of civilization from its ethical and spiritual 'core', an essential core which was always and already 'feminine'.

In this context, Partha Chatterjee's critique of Humanism in his celebrated essay *The Nationalist Resolution of the Woman's Question* (233-253) becomes significant. Here, he shows how it was precisely Enlightenment and Western Humanism that constituted the eternal 'inner' spiritual Indian core (*ghar*) as opposed to the 'materialistic' outside (*bahir*) – a binary that got reified and fossilised with time as 'naturalised' stable 'originary' categories. He also went onto depict that it was precisely the political 'truth' of Humanism which constructed and, in turn, eulogised the transcendental significance of the *bhadramahila* (a combination of *Victorian* bourgeois notions of an enlightened mother and *Brahminic* notions of wifehood) as a metaphor for the inner spiritual core of the nation, at the cost of marginalising her own lowly sisters who perhaps exhibited robust humour, sexual explicitness and frankness in their gestures and demeanour (considered too harsh or uncouth for the gentlewoman's ears). And, it is precisely this spectre of the 'fallen' lowly woman, or the 'loose' Western woman, that haunted the school's psyche, the reason why it incessantly engaged in several punitive measures in order to keep the 'essence' of (what they called) 'our *culture*' intact.

For instance, the school was heavily invested in the positioning of the *dupatta* and the manner in which the salwar was worn. We were repeatedly told to not sit cross-legged, or with legs apart, and unlike Carmel, a special emphasis was placed on Bengali so that students spoke the language during informal conversations. It also prescribed a culture of 'self-criticality' and 'self-refelxivity' owing to what they called the 'civilizational legacy' of our country. But, what happened was that this notion of 'self-reflexivity' very easily got translated into the idea of 'self-surveillance' where the girls were schooled in the idea of gentleness and sobriety. Thus, this school too, just like the previous one, had its own way of 'recruiting individuals as subjects', placing sexuality at the interstice of *anatomopower* and *biopower*[3], thereby seizing the 'body' itself, with all its materiality, through a 'studied manipulation of the individual', in an ardent spree of manufacturing 'good' elite women of the *future*.

IV. Effacement of the 'Woman'

Right from the colonial period till today, the definition of a 'good' society (the realm of what we know as 'traditional *high* culture') was realised and *materialised* in and through the body of the woman (through processes of vigorous protectionism). While imperialist discourses of the kind Patha Bhavan was apprehensive about tries to create a 'good' society

[3] With the advent of Enlightenment rationality, there was the emergence of certain techniques of power centering on the individual body, which meticulously looked into the spatial distribution and organisation of individual bodies (their allignment, serialisation etc), something that Foucault calls 'anatomopolitics', where there was certainly an emphasis on the visibility of the individual, in order to control them, render them usable and manipulable, and even increase their productive force through the tactics of exercising, drilling, etc. Now, what is significant here is that without such a regime of discipline, freedom does not get its definition, where the sense of freedom of the subject became possible in and through the proper exercise of power. Such was the internal topology of 'truth' that fed into the notion of what we know as Biopolitics, something that certainly sprung within the grid of disciplinary power itself.

by eulogising the image of a *sexy*, elite yet a respectable woman (someone who supposedly 'chooses' to be what she is, a 'free' subject), *civilizational* arguments of the kind Patha Bhavan adhered to say that *true* 'freedom' lies in being materialist in the public sphere, only to the extent that one holds on to the 'spiritual' core of our traditional high culture. And, most interestingly, both arguments engage in the fallacy of believing in the notion of 'full *intending* presence' of a culture or a subject (in the first case, by thinking that women are self-seeking sovereign individuals beyond the mechanisms of disciplining and in the latter, by adhering to the notion of an ever-lasting eternal 'Bengali culture' which Westernisation, modernity or capitalism is thought to *harm* and *appropriate*). In these two opposing notions of 'freedom' offered by neo-liberal imperialists and civilizationalists of the kind I experienced at Patha Bhavan, 'the place of the female subject is *different*', a word that Gayatri Chakravorty Spivak uses to show the position of the 'displaced' female subject while talking about the colonial debates over widow immolation or *Sati* (where apparently pro-Sati and anti-Sati positions legitimised each other). She says, *"Between patriarchy and imperialism, subject-constitution and object-formation, the figure of the woman disappears, not into a pristine nothingness, but into a violent shuttling which is the displaced figuration of 'the third world woman' caught between tradition and modernisation"* (128). Her words resonate with the manner in which the 'normative' was constituted inside the school. In spite of Spivak's disagreements with Foucault, the Foucauldian ethical impulse, which would help enhance my argument here, is the fact that this 'identity' of the *bhadramahila* which Patha Bhavan tried to construct, an unchanging reified 'identity', was certainly an 'identity-in-difference', marked by the *unfinishable* nature of an apparently 'pure' entity and a lack of *closure* or finitude.

V. Cohabitation of the Archaic and the Modern

In this respect, another point becomes extremely significant. Along with its emphasis on the notion of an eternal 'core' or 'essence', what I also experienced in the school was a simultaneous emphasis on the notion of material efficiency, modernity and progress. This apparently incommensurate cohabitation of the archaic and the modern (that is, the material and the spiritual) within the same space demands some exploration.

It is important to note that certain postcolonial thinkers adhere to the notion that while earlier, the lifestyle politics of nationalists[4] was essentially 'inclusive', the later thrust to globalise and Westernise is 'essentially' based on individualism and consumption. Lifestyle politics based on supposed 'looseness' and 'forwardness' is 'symptomatically' interpreted as a significant counterpart of 'economic globalisation', devoid of the kind of 'self-reflexivity' that the legacy of civilization taught us (*From Civilization to Globalisation* 150). This was certainly the position that Patha Bhavan adhered to in terms of disciplinary mechanisms which its students were subjected to. Thus, the school was heavily devoted to the project of 'saving' or recuperating what we now know as 'Modernity's Other', the essential 'core' of 'our' culture. But, most crucially, this nostalgia and resultant zeal for an 'essence' was certainly not only an orientalist/nationalist foundationalism for an 'originary' notion of 'Bengaliness'. Rather, this 'search for an origin', implemented in and through vigorous protectionism, should be read semantically as an ardent spree to hold on to an *ever-receding* 'signified', 'origin' or 'core'. Hence, although apparently it was an essentialist orientalist phallocentric project, it basically emanated from a sense of 'anxiety' about the *absence* of a pure authentic origin (or the presence of an *ever-receding* origin) – an origin which they thought was necessary to 'keep' in the face of the onslaught of a Westernised mode of education which the school had to otherwise adopt in terms of its method and language.

Interestingly, in this case, the archaic (the notion of the ever-lasting 'core') was actually engaged in an active dialogue with the modern, devoid of any ontological *logocentric* valency. Thus, here the analyst's job is to decode the structure to see how the archaic can be alive and kicking inside the heart of the modern itself, and thus, how the 'identity' of the student was formed in and through the category of 'difference'. Hence, such a Structuralist zeal on the part of the school to connect apparently disparate aspects of our history (notions of scientific materialism as well as that of an essential 'core') would actually act as a critique of something that it itself adhered to – the idea of an organic 'civilizational origin' of Bengali culture. Thus, this almost '*hauntological*' continuation of certain 'residual' beliefs in the 'modern' disciplining practices of the school would certainly not be read as an ethnocentric orientalist teasing out of

[4] Here I am refering to Dipesh Chakrabarty's argument in "*From civilization to globalisation: the 'West' as a shifting signifier in Indian modernity*" where as instances he mentions how Swami Vivekananda's attire in Chicago was informed by the suggestion of American women or Gandhi's choice of loin cloth was an effect of months of preparation and planning (2012).

ontological elements of tradition within modernity. Instead, it requires a *symptomatic* reading of how certain cultural scripts continue to exist long beyond their historical 'origin' by becoming codified through constant 'repetition' and thus emphasise the '*structural* aspects of a particular culture' (Chakrabarty).

VI. *Heterotopia* and the Third Space of Recalcitrance

Hence, such a theoretical matrix would not read the coexistence of Western materialism in terms of the school's language and method and its patriarchal moral vanguardism under the rubric of oppositional essences of 'Westernisation' and 'civilizational core', but would refer to the notion of a Foucauldian *heterotopic* culture of our time, which juxtaposes, in a single real place, several spaces and several sites that are in themselves incompatible" (Foucault 18). Here, the archaic is certainly not part of our 'essential' civilizational legacy, which demands a eulogy. Rather, seen from this framework, the violence of the 'logos' that a civilizational legacy carries with it got exposed instead of being deemed a more ethical, democratic and spiritual space.[5]

Such a 'polytemporal' notion of 'time'[6] helps us engage in a vigorous critique of the homogenous linearity of time as progressive since what happened in Patha Bhavan was a tantalising mingling of heterogeneous epochs, genres and ideas, thereby referring to the coexistence of 'different *non*-ontologies', where the past remains in the modern, but certainly not as an ontological 'return of the repressed'.[7]

Most significantly, in this heterotopic space inside the campus, there lay a 'third space', between a law informed by discipinary power and biopower constructing the students on one hand and on the other hand,

[5] Such a heterotopic space certainly resonates Foucault' words, where he says that, "the notion of heterotopia marks "the will to enclose in one place all times, all epochs, all forms, all tastes, the idea of constituting the place of all times that is itself outside of time and inaccessible to its ravages, [...] this whole idea belongs to our *modernity*" (Foucault 1967).

[6] Bruno Latour (1993) uses this term in *We were Never Modern*.

[7] For Foucault, the narrative of 'time' as a longitudinal category gets *constituted* with modernity, comes within the realm of the *thinkable*, where our imagination gets arranged diachronically, and 'History' becomes the dominant mode of analysis. Space gets arranged according to time (resonating his analyses of the time table at the school, the prison or the battalion), where the body is rendered docile, i.e. usable and manipulable, through the cultivated syncronisation of the body penetrated in time, in such a manner that space cannot be conceived without the grid of vertical temporality.

another law which is 'illimitable' and always surpassing those who aim to control it (Golder and Fitzpatrick 55). It is this 'irreducible' *recalcitrant* space which the students siezed upon to engage in several transgressive and resistant practices against certain prescribed modes of conduct – transgressions which were a result of the 'subjectifying' and 'liberating' practices of the *norm*. Thus, most crucially, the *internal coherence*, 'positivity' and 'truth-making' practices of a normative discourse were certainly punctuated by the notion of transgression. It is precisely these moments of transgression which brought into being the 'positivity' of the hegemonic discourse of the 'gentle elite woman', something that also hinted at the *incomplete* or *inadequately complete* nature of power, a power that is forever lacking in the all-so-important notion of a secure 'centre' of an ever-present fiction of 'Bengaliness' that the school desperately tried to achieve through its normalising devices, only to hint at its vacuity.

However, the discourses which tried to appropriate the students at the same time also opened up possibilities for *resignification* that was ever-exceeding in nature. Here, the students were called upon to be at constant war in order to 'secure' *our heritage* against Western ailments, against *rogues*[8], marked by any element of non-conformity and disorder. Such a 'civilizing mission', an act of 'recovery' instituted as a response to the perceived *ill-effects* of a 'foreign' culture (which was otherewise adopted in the curriculum), should be construed as an act of 'poesis', not in a romantic sense, but rather as a *technique*, of bringing things into being, 'naming', creating a mode of 'differentiation' and making meaning, enframing and proliferating. Thus, this act of 'recovery' should not only be perceived as a disciplinary act which was merely instrumental, where students were treated as mere 'cases' through governmentality, but also an act of imagination related to the notion of 'freedom' that rearranged metaphors and brought a new language into being. Hence, in this case, discipline ceased to be a mere *centripetal* process, one which prescribes, concentrates, forbids and encloses, but rather became a *centrifugal* mechanism of regulating 'life' itself (Dillon 16). Here, the disciplined student in the school is not opposed to the notion of the 'liberated' student, since the disciplined self-surveilling gentlewoman is the *result* of liberties.

Most significantly, in order to promote the notion of the 'good respectable (female) student', disciplinary practices must first 'know' her –

---

[8] The concept of the rogue and of roguery is derived from early modern theories of biology, where Charles Darwin (1959) in *The Origin of Species* referred to 'roguing' as the practice by which plants that deviated from the 'proper' standard of their species in seed-beds were weeded out.

precisely why there was an emphasis on one-on-one interaction with teachers and creation of an environment where students would not merely be mute listeners but 'speak out'. Thus, precisely through this negation of what Foucault called 'repressive hypothesis', what individual trait requires punishment and 'correction' with the goal of preserving the larger bio-political body was determined. Thus, the very processes which aimed to produce and generate hegemony signified also sought to kill and demolish all the aspects which were considered inimical to the preservation or recovery of that signified. More crucially, this 'non-dilectisable antinomy'[9] (a kind of an interstice where incompatible forces struggle against each other, devoid of any *dialectics* or *synthesis*) between the mechanisms of disciplining and the ever-exceeding recalcitrance also hinted at the event of the interruptive, undecidable *decision* – what Foucault perhaps hinted at through his notions of the 'Singularity of an Event' or 'Emergence'[10], where the element of the 'political' was certainly recuperated, where metaphysics was constituted by the traces of its own deconstruction within itself, since any discourse of the *bhadramahila* was *always and already* haunted by the spectre of the 'fallen loose Westernised woman' too.

VII. Normalising Power and the 'Gift' of Subjectivity

Significantly, while Patha Bhavan engaged in a set of *tactics* that aimed at 'writing' the future of its students in a particular way, predicated on its idea of the 'right' combination of the spiritual and the material, those very tactics helped me to write the same future in a different way perhaps – precisely why I could endeavour to write this article. The very fact that I can interrogate my vigorous schooling experience comprising 16 long years is certainly not only due to my training in the humanities, but rather is a result of a process which happened in and through the effects of normalising power. According to Foucault, 'Wherever there is power, there is resistance'. Here, the *normalising power* of Patha Bhavan acted as the Derridian notion of 'gift' (of subjectivity) for me, something that is certainly not perceived as a liberal self-fashioning performative mechanism, but rather a kind of normalisation which refers to the prior notion of poesis or 'dissemination' of variant *subjectivities* (intertwined with the idea of the *techne*), where normalising power is considered one that 'never returns to itself', a power which *constitutes* but can never

[9] Derrida uses this term in "Rogues"(2005).

[10] *Enstehung* or the "moment of arising" (Emergence) is something that designates a place of confrontation amongst incompatible forces which always occurs in the interstice, amidst an endless play of dominations.

determine. Such a notion of an 'illimitable' power as dissemination also gave rise to the *'unthinkable' of the 'sameness' of the episteme*, i.e. the 'gift', where 'resistance' on the part of the students was not an 'originary' act but rather a reconfiguration and thus a resultant disruption of the circulation of normalising power.[11]

Every repetition was a rupture here in some sense. Here, the 'self' certainly became that open topos of ethical encounter – plural and divided, something that I perhaps would have missed out on had I not experienced the techniques of such vigorous mechanisms of normalising power at that time. Furthermore, it is precisely because I did not take the notion of the 'spiritual core' as a self-evident 'natural' category which had to be secured, and instead considered it 'imagined' and 'constituted' by constructing the discourse of the dangerous 'fallen Other', I was later able to *reimagine* and *reconstitute* it by construing the notion of 'being' as an enterprise, a project, and certainly not a *finished* product bearing the marks of an eternal fixity.[12] And, in the process, I was also capable of engaging in minute 'descriptions' of the concrete facticities in order to critique the transcendence of 'Being' and yet not deny its presence in the everyday, and show how the *Being* (universal) is realised in and through the *beings* (concrete),[13] and not the other way round (Derrida 190).

VIII. Opposed 'Essences', Same *Episteme*

In one of his accounts, Rabindranath Tagore articulated his deep-seated despair regarding formal school education based on his experience in the schools that he went to, the last of which was St Xavier's Collegiate School. He wrote, *"But I can never connect myself with a school that shuts out life and natural beauty and conjures frightening images of a prison or hospital. We had to sit like dead specimens of some museum while lessons*

---

[11] Here, normalising power is certainly considered to be a disseminatory force, thereby giving rise to the Derridian notion of the gift, which refers to a form of 'giving' gift/power where the giver is not conscious of it, neither is the sender, and moreover, the gift given/power disseminated is also not recognised as such (Derrida 1987).

[12] However, here I do not claim 'transcendence'. Instead I aim to hint at the political imperative of being 'hopeless' about the totalising tendencies of disciplinary power, certainly not in the sense of lament, but as a critique of teleology.

[13] Here, Heidegger's notion of "Being-With", a kind of an anti-subjectivist community, where every Being is a Being-With, devoid of the Self and the Other binary, becomes significant (Grosz 1997).

were pelted on us from high like hail storms on flowers"[14] (Chakraborty 214).[15]

Ironically, after spending two years at Patha Bhavan, I myself got admitted to St Xavier's College itself for my Bachelors (a place which had its own pedagogic processes of humanisation, new sets of departures interspersed with novel continuums), an institution whose methods were vehemently critiqued by Tagore – precisely why years later he aimed to realise his dreams of opening up a comparatively 'liberating' space by instituting Patha Bhavan at Visva Bharati. The question, in this case, is not whether the school in Calcutta was able to evince a similar spirit of supposedly 'free and unrestricted thinking' that was practiced at Shantiniketan. I was relieved to leave Patha Bhavan since the apparently 'informal' way of learning by staying close to the 'natural environment' certainly did not augur well for me. Moreover, it taught me how both 'Naturalism' and 'Social Constructivism'[16], two apparently contradictory claims to knowledge, in actual practice, interact and refer to each other within the *same* epistemological network and hence, it is foolhardy to assume that 'nature' and 'rawness' (epitomised by Patha Bhavan) revealed 'truth' while 'society' and 'fabrication' (epitomised by Carmel) revealed 'errors', since such apparent binaries like 'truth' and 'error' are all part of the same *limit of the thinkable*. Such a realisation also helped me do away with the 'depth' question, and the notion of *false consciousness* as well. Moreover, such varied *experiences* (and I certainly use it for want of a better word) indeed helped me to engage in a vigorous critique of the notion of 'origin' and an 'aunthentic self-seeking subject', since each discursive framework has its own organising logic. Thus, we certainly encounter a double bind here. We can call it 'Foucault's Fork'[17] for the time being – a tricky fork which shows how each institution has its own ways of disciplining, its own ways of 'making-live' and 'letting-die', where apparently incommensurate and antithetical 'essences' create the same *logos* and thus belong to the same *episteme*.

---

[14] See Chakraborty, Amiya (ed.)"Tagore Reader", Macmillan & Co.,London 1961, pg. 214.

[15] Tagore further goes on to say, "The rooms were cruelly dismal with their walls on ground like police man. The houses were more like a pigeonholed box than a human habitation. No decoration, no picture, not a touch of colour, not an attempt to attract the child's mind." (See Tagore, Rabindranath, "Reminicences", Visva-Bharati, 1961, pg. 60-61.)

[16] *Naturalism* is where society is immanent and nature is transcendent and *Social Constructivism* is where society is transcendent and nature is immanent (Latour 1993).

[17] A term coined by Ian Hacking.

Works Cited

Althusser, Louis. "Ideology and Ideological State Apparatuses", *Lenin and Philosophy and Other Essays*. Trans. Ben Brewster. New York and London: Monthly Review P, 1989. 170-186. Print.

Bhabha, Homi. "Signs Taken for Wonders: Questions of Ambivalence and Authority under a Tree outside Delhi, May 1817" and "Of Mimicry and Men: The Ambivalence of Colonial Discourse". *The Location of Culture.* London: Routledge, 1994. Print.

Chakrabarty, Dipesh. "From civilization to globalisation: the 'West' as a shifting signifier in Indian modernity". *Inter-Asia Cultural Studies*. 13.1(1 March 2012): 138-152. Print.

—. "Subaltern Studies in Retrospect and Reminiscence". *Economic and Political Weekly*. XLVIII.12 (March 23, 2013). Print.

Chakraborty, Amiya, ed. "Tagore Reader". Macmillan & Co., London, 1961. 214. Print.

Chatterjee, Partha. "Anderson's Utopia". *Diacritics*, 29.4 (Winter 1999). 128-134. Print.

—. "The Nationalist Resolution of the Women's Question". *Recasting Women: Essays in Colonial History.* Ed. K.Sangari and S. Vaid. New Delhi: Kali for Women, 1989.233-253. Print.

Derrida, Jacques. "Women in the Beehive: A Seminar". *Men in Feminism*. Ed. Alice Jardine and Paul Smith. New York and London: Methuen, 1987.189-203. Print.

Dillon, Michael. "Must We Secure Security?" *Politics of Security: Towards a Political Philosophy of Continental Thought.* New York and London: Routledge, 1996. 14 – 22. Print.

Foucault, Michel. "Course Summary", 'Lecture 3', 'Lecture 8' and 'Lecture 11'. *Society Must Be Defended: lectures at the Collège de France, 1974 – 1975.* New York: Picador, 2003. 43 – 64, 167 – 188 and 239 – 264. Print.

—. "*Discipline and Punish: The Birth of the Prison*". Trans. Alan Sheridan. New York: Random House, 1975. Print.

—. "Of Other Spaces", *Heterotopias*. French journal Architecture/ Mouvement/Continuité, 1967. 1-20. Print.

—. *History of Sexuality*, Vol. 1. New York: Vintage Books, 1980. 135 – 159. Print.

Golder, Ben and Fitzpatrick, Peter. "Foucault's Other Law", *Foucault's Law.* New York and London: Routledge, 2009. 53 – 98. Print.

Latour, Bruno. *We Have Never Been Modern.* Cambridge MA: Harvard UP, 1993. Print.

Spivak, Gayatri Chakravorty. "Can the Subaltern Speak? Speculations on Widow Sacrifice". *Wedge*. 7-8 (Winter/Spring 1985): 120-130. Print.
Tagore, Rabindranath, "Reminicences". *Visva-Bharati*, 1961. 60-61. Print.

CHAPTER FOURTEEN

IDEOLOGY OF THE LIPS: FEMININE DESIRE, POLITICS OF IMAGES AND METAPHORIZATION OF BODY IN GLOBAL CONSUMERISM

SAMRAT SENGUPTA

The objective of this essay is to locate the notion of the 'feminine' in the context of the current neo-liberal world economy. While there can be too many texts and contexts of feminist scholarship that interrogate and measure the notion of the 'feminine', this essay will primarily focus on contemporary visual representations of the same in the background of a certain commodity culture. While some feminist activism and scholarship focused on positing a certain equality among genders and, in a larger way, claimed equal status and rights as men, others have questioned the ethics and politics of gendering.

Epistemic enquiry of values would show that whatever is valued and measured as positive is usually masculine or male in principal. So, the investment of certain strands of feminist scholarship has been to bring to the fore a logic of feminine as against the masculine construction of reality and point out the limitations of the masculine principals and bring back the 'feminine' as a positive quality in thinking and doing things. The early phase of feminism tried to expose how the manmade world keeps women away from certain operations which remain exclusive to men. By making those activities inaccessible to women, they are disempowered by the power structure. Society almost organises things in a way that women are forced to take a secondary, marginalised position vis-à-vis men. So, the possibility of emancipation of women is located in jeopardising that naturalised order and taking part in the work of the world as equals of men. This is the standard liberalist argument of some feminists.

The poststructuralist turn, on the other hand, has enabled us to rethink the positions of superiority and inferiority, centre and margin or masculinity and femininity. If these positions are predesigned, then moving from one to another would not necessarily put an end to structures of domination – it would, rather, produce new forms of hierarchy and marginalisation. Then, it is important to intervene in the knowledge system to show that the way we make sense of reality is itself oppressive, authoritative and gendered. This gendered reality is what Derrida would call phallagocentric i.e. which has one central truth and is overdetermined by the patriarchal logic of the centrality of the phallus. Just as in the case of gendering, the presence or absence of the phallus marks the identity of a person as male or female and likewise his/her value is determined in the society, similarly, at the level of discourse and representation, certain behaviors or modes are privileged over others as better or closer to truth.

Logically, the valued ideas and signs are predominantly masculine by social standards and therefore dominant and superior. In a larger way, our thinking preempts a certain parallel to the gendering of the being. But the question is whether it is important to dismantle this structure of thinking. Is it possible to locate and then dislocate sexual difference? If we are locating sexual difference in human thinking with exactness, we are fixing such difference and reducing it to the logic of presence/absence or self/other. In this logic, then, there can be only one gender – the other one is simply the absence of it. There can be no sexual difference then. The moment of location of sexual difference becomes the moment of its dislocation. The opposite movement is also true i.e. if one is trying to outdo/dislocate sexual difference and achieve neutrality, that neutrality must again get inscribed in the logic of phallocentrism by fixing up the neutral. Derrida writes: "One ensures phallocentric mastery under the cover of neutralisation every time" (Derrida 1992 101). Going back to the neutral sexually undifferentiated imagined stage is to destroy the other gender – the other being, the feminine, in our phallocentric world of habitation. It becomes similar to the liberalist understanding of feminist emancipation – becoming equal to man by being capable of taking his position.

Derrida writes: "Such phallocentrism adorns itself now and then, here and there, with an appendix: A certain kind of feminism" (Derrida 1992 101). On the other hand, making sexual difference absolutely locatable and therefore celebrating the specificity of a certain notion of femininity is also to fall into the epistemic trap of essentialising sexual difference and thereby submitting to the masculine understanding of the world in terms of hierarchy and superiority – simply replacing and reversing masculine superiority with feminine. This produces the aporia of the feminist project of questioning patriarchy – its possibility is already stroked out by its impossibility. In the context of representation of women in a visual field like pornography, Robin Ann Sheets writes: "The problem is clear… if feminists accept such 'perceptual essentialism' and agree that all forms of representation are harmful to women, they will lose the ability to communicate, relinquish the right to represent their own sexuality, and deny themselves pleasure" (Ann Sheets 640). So, if both what a woman wants to be or what she desires and what she is are manufactured in accordance with patriarchal episteme, how will one address the question of female autonomy, choice and the question of confronting the negation of femininity as absence of the positivities of maleness?

One possible way of addressing this paradox is to turn the structure inside out – to speak from within the structure about its possibilities which can't be accommodated by its own rational structure – which is already there in the structure but goes beyond the logic/logos/phallus of this structure. If women are attributed a position opposite to the reason/logic/structure of patriarchy – a position of otherness – then, this otherness might not be simply the other of a self, a subject, a truth which can be understood and the absence of which constitutes the feminine other. It is, rather, the otherness which constitutes the mark of the self, its signifier. The self is already marked by its difference from other beings and selves. The self is already marked by otherness in this regard – by its own peculiarities and particularities. The otherness, then, in the logic of Levinas, can be thought of as the other of the other – the radical alterity which cannot be accommodated with the logos of the self/the being. If it could have been done, there would have been no other of the self, no woman who is not man, no absence that is not presence. Feminists since Julia Mitchell have turned down the claim of some earlier feminists to read Freud absolutely in terms of patriarchal bias for his remarks like 'a woman's mind is a dark continent' or 'it is not possible to know a woman'. With the shift of scholarship from hermeneutics to phenomenology and, hence, structuralism and post-structuralism, what a woman wants becomes a non-question – it, rather, becomes important to understand femininity itself as a phenomenological void within the structures of patriarchy. However, this void is not considered debilitating by feminist scholars like Irigaray. Irigaray comments: "'She' is indefinitely other in herself. This is doubtless why she is said to be whimsical, incomprehensible, agitated, and capricious… not to mention her language, in which 'she' sets off in all directions leaving 'him' unable to discern the coherence of any meaning. Hers are contradictory words, somewhat mad from the standpoint of reason, inaudible for whoever listens to them with readymade grids, with a fully elaborated code in hand. For, in what she says, too, at least when she dares, woman is constantly touching herself." (Irigaray 28-29)

However, Irigaray's position is different from Levinas' though her conceptualisation of the feminine other as alterity to the maleness of reason is taken from him. While Levinas would refer to 'the face of the other' as recognition of alterity for the self and think of the feminine other as an ethical absolute like God who is ungraspable and is placed vertically above. While Levinas thinks of an Other (with a capital 'O'), Irigaray would resist any such possibility of visual representation of alterity – of the unrepresentable woman in the discursive speculum of man. The manmade visual world, Irigaray points out, is already operated by a logic of what she ironically calls specula(risa)tion (Irigaray 177). The reflection against a speculum is an act of speculation of the male subject. Irigaray writes: "in order to reflect (oneself), to speculate (oneself), it is necessary to be a 'subject'" (Irigaray 177). Irigaray would therefore consciously resist any attempt at identifying the other as anything coherent that can be represented visually and morphologically as complete. She would prefer touching over seeing. Feminine pleasure is more in touching than in seeing. Irigaray here displaces the overbearing metaphor of penis/phallus in Western thinking, particularly in psychoanalysis, with the image of lips which is divided and is two; furthermore, it is defined by an interruption and touching of each other as opposed to the continuity of presence and determinism of phallus. Irigaray writes: "that contact of *at least two* (lips) which keeps woman in touch with herself, but without any possibility of distinguishing what is touching from what is touched" (Irigaray 26). Touching is like the Derridean metaphor of the 'blink of an eye' as opposed to continuity of vision/presence/truth that dominates Western metaphysics from that of the Greek period. Irigaray writes: "her sexual organ represents *the horror of nothing to see*... It is already evident in Greek statuary that this nothing-to-see has to be excluded, rejected, from such a scene of representation." (Irigaray 26)

In Indian popular media, for a long time, desire has always been featured as unarguably male, women being the object of desire. In printed ads, ad films and feature films, women's bodies have always been portrayed as objects of desire and associated with goods which are consumable. However, after economic liberalisation, when the expansion of the market became inevitable, and in a service-industry based economy where more and more women were becoming autonomous and the argument of the freedom of making choices became crucial for both the sexes, feminine desire and its portrayal became a major preoccupation. Thinking along the lines of the above debates, we see, epistemically, how women's desire itself is constituted and framed by patriarchy. The notion of what is feminine is a part of that episteme. So, the question that can be put forth is whether the portrayal of women's desire and desire that is 'feminine' in principal leads towards their autonomy. The 1995 Bollywood film, *Dilwale Dulhaniya Le Jayenge (The Big-hearted Will Take the Bride)*, featuring the extremely popular hero, Shah Rukh Khan, contains a song 'Mere Khwabo Mein Jo Aayein' (The one who appears in my dreams) that shows how the heroine Kajol is fantasising about her dream lover whom she is yet to meet in the film. This particular sequence appears near the beginning of the film but we see that the hero already appears as a concrete visual entity in the dream of the heroine, even before he appears in person in the movie. The plot of this film revolves round the theme of how Kajol opposes her father's wishes in rejecting the groom he has already selected for her and continues her quest for a dream lover after she meets Shah Rukh Khan and asserts her autonomy and free will against the structures of the society she lives in. Maybe in post-liberalisation India, she represents the true global citizen – the new woman who can assert her choice against the forms of tradition decided for her by family and society.

The question of choice becomes important in the context where we see Kajol playing Simran who comes from a middle class conservative north Indian family and goes on a tour of Europe where she meets Raj, the hero of the film, played by Shah Rukh Khan. He is financially stable and settled in London. Critics have argued about the transformation of the image of the hero from one who is struggling to establish himself and prove his worth to an intensely consumerist hero, economically well endowed by an affluent father, who "has no history for he is a product of the liberalised market" (Bakshi and Sengupta 45). The rise of middle class and its immersion in the new culture of consumerism become evident as we see Simran having the time of her life – fulfilling her wishes in Europe, outside the barriers and restrictions of her family. However, in the beginning of the film, we can see her choices already being established. The physically active, playing, swimming, running hero, Shah Rukh, appears in her dream beforehand. On the other hand, she also becomes a desired object – one who has subjectivity, who, instead of being passively consumed, desires consumption – the new woman who is fit for the neo-liberal world order. If autonomy is not manufactured for the subject, how can he/she become a potential buyer? The clean and clear division of subject and object collapses in a consumerist economy. Mary Louise Roberts, in an essay, comments: "In the 'specularised' urban culture of arcades, boulevards and department stores, woman was inscribed as both consumer and commodity, purchaser and purchase, buyer and bought"

(Roberts 818). It is important to remember here how Irigaray asserted feminine pleasure in touching "without any possibility of distinguishing what is touching from what is touched". The problem with such dramatised equality of what is touching and what is touched lies in the very act of their naming. Naming gives identity and identity sets the limits of becoming. It is not enough, therefore, to ascribe subjectivity to women – give them the chance to choose, select and possess because, in the very staging of sexual difference, the subjectivity of women is already objectified. Here, they become subjects only to get objectified – the subjectivity gets codified into a fixity which is then desired.

Here, I would like to refer to a couple of advertisements. In the first one, which is that of a body spray, Wild Stone, we see how a woman clad in traditional attire, a sari, during an auspicious festival like Durga Puja, gets attracted to a man because of the aphrodisiac fragrance of his perfume and gets into bed with him for a quick fling. The man in the advertisement is the actor, John Abraham, who is extremely well built, with a good physique and conforms to the global construction of masculine beauty. In this ad, the woman's attraction towards the object – the fragrance/the body makes her desirable. We see in the bed sequence that the man is on top pleasuring himself. The target audience of this body spray is, ironically, men who would locate female desire in it. Now, who is consuming whom here – the woman consuming the man, the man consuming the woman or both consuming the product and in turn getting consumed by it? The other ad which I would like to discuss shows all objects, including the navels of girls, pizzas and the garage, having lips as the actor (Shah Rukh Khan again) buys and drinks Pepsi. If women are ascribed subjectivity, it might only be to make them objects in turn. Desire in the neo-liberal world is not simply to consume but to get consumed by that desire for consumption. This is how desire can endlessly reproduce, reflect and relocate itself. This is what the market wants – desire as a contingent ephemeral thing like touching, like the blink of an eye – every moment of consumption becomes a fleeting moment and keeps one waiting for the desire to arrive. The desire becomes the name of the desire. Feminine alterity, multiplicity of otherness then get marked by the logos of desire which is, however, a non-logos – alterity itself, which is divided, many and infinitely exchangeable without reason.

This takes us back to the old paradox of pornographic representation of women – on one hand, it disentangles the body, woman's body, her pleasure from silence and non-representation but on the other hand, it objectifies her – commodifies her for the sake of men's pleasure. In the following pornographic image, where one woman is seen with several men, it is again difficult to identify the giver and taker of pleasure. The woman stands at the centre – around her, the men organise themselves.

The myth of vagina dentata or toothed vagina is a part of human culture where the castrating phallic mother threatens to devour the phallus – tear it off. It is a part of patronymic fantasy which still renders the positive power of the phallus. The absence of the phallus causes anxiety of castration in men, rendering them limited autonomy in the patronymic power structure which might be taken away. The powerlessness of women comes back ironically as an apparent empowerment when it makes man realise how he might be pushed into a similar condition of powerlessness – dislocated from the phallic centre of truth. However, this keeps the patriarchal myth of the existence of a phallic centre alive and represents femininity as lack of it. The lack itself, then, becomes namable and the gap between two lips is given an identity – an identity as an interruption, a non-being, a hole. This name is marked with the name of the phallus in terms of its lack. If there is a lack, it marks the absence of a presence. It is true that this absence partly gives identity to the presence, the logos/phallus/truth, but is itself marked by it. The vagina is marked by the anxiety of penetration. This kind of image, this kind of imagination reproduces the structure of production-consumption, hermeneutic unveiling, tearing of hymen, placing the phallus inside the gap. The consuming subject is also the object to be consumed by the object of consumption. We have already seen that. The role of subject-object can be interchangeable. The privileging of the element of feminine in deconstructive feminism is received by Gayatri Chakraborty Spivak with deep skepticism. Making women the 'subject' of deconstruction has its double bind. Spivak writes: "We give the 'subject' its philosophical value of the capital I... But colloquially, 'my subject' means 'my object'" (Spivak 1997 46). So, giving identity to the lack, making alterity/femininity/object/thing-in-itself the 'subject', can, in a reverse move, make it the object. The danger of such reversal is twofold – on one hand, it might reproduce the myth of vagina dentata – the castrating mother – fetishising the otherness in woman as empowering and on the other hand, might essentialise femininity as lack of masculine aggression and violence, thereby producing the myth of sacrificing motherhood. Both global consumerist culture and religion objectify – they not only visualise subjects as objects but also render subjectivity to objects – the subjectivity ascribed is, however, imbued by the threat of castration – threat of being removed anytime.

Just like feminist champions of pornography or liberal feminists pointing out how market economy has enabled visibilisation of women's freedom, the divine feminists would argue for the mother goddess figure as being empowering for women. The opposite camp would, however, posit divinity as manmade and therefore subject to a certain objectification. The divine mother is the one who can contain the radical alterity – the unnamable – she becomes the act of naming the unnamable – the alterity which can be posited by man – put to use within the manmade oikonomos. Spivak, in her essay, 'Moving Devi', writes, "The many representations of the great goddess look stunning on the wall. Real women are distanced from her. She is no role model unless... one of us is thrust into that space" (Spivak 2008 207). The capacity of *dvaita* (dualism) to make possible the descending/the *avatarana* of the outside, the other in the ordinary, the everyday becomes a taking away of subjectivity of the woman who is pushed to that position of *devi* – the divine mother goddess. Such is the use of alterity which is named and given a unity. Similarly, the consumerist economy gives a unity to the critical charge of alterity. If divination makes alterity usable, consumerism makes it exchangeable. The blink of an eye gives every moment a new vision, a new object to be consumed. If the 'other' becomes namable, if the 'feminine' becomes 'the subject' and acquires an identity, it becomes phallocentric like the phallic, castrating mother. Maggie Berg comments on an acute misreading of Irigaray's position on feminine alterity: "To assert woman's difference, as some critics claim Irigaray does, would simply make the lips the new phallus… In Lacan's case, ironically, the phallus turns out to be the 'empty' womb: The contentless origin or matrix of the Symbolic. Irigaray's lips are an alternative to Lacan's phallus but Irigaray does not constitute them as either 'univocal' or privileged in the order of 'being'; she does not pretend the lips are the 'privileged signifier' of our culture" (Berg 70). To understand the functioning of the lips as the new phallus, the feminine as the consuming name of the desire, I shall draw the example of an advertisement of Levis jeans which shows a woman wearing Levis jeans and a skimpy top lying down on a heap of male bodies which, ironically (or may be intentionally), resembles the infamous human pyramid of Abu Ghraib with the words "This is the way to rule the world" inscribed beneath. This is the truth of global consumerism – the name of the desire as non-truth. The strategy of ruling has moved way beyond the strategic essentialism of positing and sublimating the feminine as non-truth against the patriarchal, imperial, masculinist world order. Derrida posits the non-truth of alterity in the figure of hymen that is unseen – the fold between two lips. For him "woman knows that castration does not take place" (Derrida 1979 60). The capacity of the feminine 'other' is that it/she emulates castration knowing that it never happens – it/she dissimulates truth. Derrida asserts "There is no such thing as the truth of woman, but it is because of that abyssal divergence of the truth, because that untruth is <<truth>>. Woman is but one name for that untruth of truth" (Derrida 1979 51). However, Spivak is skeptical of such naming of the woman as untruth, positing her to the fullest as a namable alterity of the order of maleness, She calls it the "'feminization'" of philosophising for the male deconstructor" (Spivak 1997 52). The figure of the castrating woman in Spivak gets replaced by the woman who knows that there is no truth; there is no phallus and therefore no castration. Knowing that woman knows this but can still simulate castration and therefore this simulation is dissimulation, knowing that the real referent of the performance of castration is not there but women can dissimulate this, the male deconstructor wants to feminise his own philosophy, imitate the play of women with truth, knowing that it doesn't exist. The point is that the subject is always already an object. The ontic myth of a stable pre-existant being gets destabilised with the confrontation of dissimulating femininity. Spivak's discussion of Derrida's *Glas* and *Eperon (Spurs)* shows how Derridean reading of femininity has the potential of moving beyond the essentiality of the body metaphor. Possessing or not possessing the phallus is not that important, nor is important the position of its lack as an empowering alterity. The strategic essentialism of using femininity as a concept-metaphor for alterity to the masculinist world-order should then be replaced by a shifting strategy knowing that neither phallus, nor its lack is actually there – physically present. Spivak writes: "With her it is not a question of having or not having the phallus. She can change it, as if she had a collection of dildos or transvestite underwear" (Spivak 1997 52).

Here, the cyborg woman comes on stage – the subject which is always already the machine/thing/object – no myth of primordial possession or dispossession of truth/non-truth can define her completely.

I shall end my essay here with reference to two short stories by Nabarun Bhattacharya – a Bengali writer who constantly engages with questions of idealism, politics after neo-liberalism and fate of communism in Bengal and elsewhere. In the first story titled "Parijat and Baby K" (Bhattacharya 74-76), we see Parijat, a medical representative, meeting a prostitute called Baby K who drinks petrol in a petrol pump. Instantly, he becomes curious about her and wants to possess her. In the story, we get to know that K stands for the Bengali slang 'khanki' which means whore. Whore in the economy of this new age is an ambiguous figure. Her presence is marked by a double bind. On one hand, she is an earning individual – economically independent and autonomous; on the other hand, she is at the same time objectified – used as an object of pleasure like a commodity which can be bought or sold. She keeps reminding us constantly that the subject is always already an object. However, her being Baby K is marked by her marginality, by her being the whore she is whose body is objectified and used. Standing between subject and object, collapsing the boundary between the two, she is a shifting position that is unnamable or as Spivak would call 'a position without identity'. In Bengali, the term "K" simply means "Who?" Baby K signifies that unnamable question mark – the "who". However, this enigmatic contentlessness of the "who" can become explosive any moment. In the story, we can see Baby K drinking petrol at a petrol pump. Therefore, she is already combustible. She is not simply a victim of a world order – a certain kind of 'just natural' form of violence which can destroy her any moment like "K" in Kafka's *Trial* but, though marginalised and turned into the 'other' by a certain violence of the consumerist world order, she is combustible without intention.

In another story titled "Baby K" (Bhattacharya 77-82), we see Parijat picking up Baby K and having sex with her. After that, Parijat continuously keeps burning inside. This is the double bind of femininity –

the moment of positing/possessing her is the moment of corruption of the inside, deconstruction of the stable being, the haunting of the unnamable. Then who possesses is always already marked by what he possesses – the object is there in the subject – the subject is always already the object – the object produces the subject. The other of the being is there at the heart of it and cannot be given a separate name of non-being – it is inseparable, undecidable and haunting. The attempt of naming such undecidability might be horrendous. The circuit of consumption can explode any moment. Nabarun Bhattacharya speculates that Parijat, in that night of picking up Baby K, might push his bike till Iraq and American soldiers might, in a mood of revelry seeing Baby K, light a cigarette in her mouth and then she, with LPG inside, will turn into a human bomb – a Molotove cocktail. The lips of Baby K – the modern consumerist hybrid cyborg woman, if visible – if namable, get marked by the possibility of penetration – the haunting of the absent presence of phallus. Any moment, the namable feminine alterity can be penetrated by the unnamable phallic presence causing explosion. The human bomb turns the undecidability of death into the decidability of suicide. The play of subject-object in consumerist economy without acknowledging the presence of each in the other causes the collapse of two into one. It is deeply ironic that the human bomb resembles consumerist confusion of who consumes whom. Here, we are confused about who kills whom, who is the subject and who the object. As Derrida has repeatedly shown, sexual difference constitutes difference as such and is constituted by it (Derrida 2008 7-26); it is important to understand the gendered perception of existence itself – how the phallic penetrating airplane approaches the two lips of the twin towers – the name of neo-imperialism – the name of the undecidability of consumerist desire. Finally, it can be said that neo-liberal bio-governance has attempted to manage and name the negations of the grand narrative of civilization and progress – the women, the marginalised and the subjugated. The irony is that power itself plays the powerless – epistemically performing the counter-phallic. In the name of the service industry and consumerist economy of free-choice, neo-liberalism performs the archetype of woman ready to spread her legs – inviting consumption. By naming and signifying the non-truth of existence as such – the feminine void of unrepresentability neo-liberalism exerts free-choice and free-market as a site of masculine anxiety of the so-called non-West. The anxiety of vagina dentata or toothed vagina continues to operate. It replays the drama of sexual difference in a new way. The femme fatale that is consumerist economy invites suicidal bombers who would kill and die to put an end to the anxiety – to realise the ultimate possibility of penetrating the gap with meaning. The postmodern assertion of anything goes gets penetrated and killed by the phallus of pre-modern suicide bomber. We have to rethink how the consumerist economy of becoming more consumable already contains a theory of explosion.

Works Cited

Ann Sheets, Robin. "Pornography, Fairy Tales, and Feminism: Angela Carter's 'The Bloody Chamber'" in *Journal of the History of Sexuality.* 1. 4 (Apr., 1991):640. Print.

Bakshi, Kaustav and Samrat Sengupta. "Waking up to a Dream: Contemporary Bollywood, the Yuppie Shah Rukh Khan and the Great Urban Indian Middle Class". *Journal of humanities and social sciences.* 6 (August 2009): 41-53. Print.

Berg, Maggie. "Luce Irigaray's "Contradictions": Poststructuralism and Feminism" in *Signs,* 17. 1 (Autumn, 1991): 50-70. Print.

Bhattacharya, Nabarun. *Prem O Pagol (Love and Madness).* Srirampur: Saptarshi Prakashani, 2007. Print.

Derrida, Jacques. *Spur.* Chicago and London: U of Chicago P, 1979. Print.

—. "Geschlecht 1: Sexual Difference, Ontological Difference". *Psyche: Inventions of the Other*. Ed. Peggy Kamuf and Elizabeth G. Rottenberg, Stanford: Standord U P, 2008. 7-26. Print.

—. "Choreographies". *Points: Interviews, 1974-1994.* Stanford, California: Stanford U P, 1992. 89-108. Print.

Dilwale Dulhaniya Le Jayenge [*The Big-hearted Will Take the Bride*]. Dir. Aditya Chopra. India, October 1995. Film.

Irigaray, Luce. *The Sex Which is not One*. Ithaca, New York: Cornell U P, 1985. Print.

Roberts, Mary Louise. "Gender, Consumption, and Commodity Culture". *The American Historical Review.* 103. 3 (Jun., 1998): 817-844. Print.

Spivak, Gayatri Chakraborty. "Displacement and the Discourse of Woman". *Feminist Interpretations of Jacques Derrida.* Ed. Nancy J. Holland. Pennsylvania: Pennsylvania State U P, 1997. 43-72. Print.

—. "Moving Devi – 1997: The Non-Resident and the Expatriate". *Other Asias.* Oxford: Blackwell Publishing, 2008. 175-208. Print.

CHAPTER FIFTEEN

CORPORATE GLOBALISATION AND ITS IMPACT ON WOMEN IN INDIA

M MOHIBUL HAQUE

> It is true, or at least theoretically possible, that there are times in the life of a people or a nation when the political climate demands that we – even the most sophisticated of us – overtly take sides? I believe that such times are upon us. And I believe that in the coming years, intellectuals and artists will be called upon to take sides, and this time, unlike the struggle for Independence, we won't have the luxury of fighting a 'colonising enemy'. We'll be fighting ourselves. (Roy 198)

The universalisation of capitalism through liberalisation and privatisation is smartly projected and innocently accepted as globalisation. It works through multinational corporations and international trade and financial institutions operating in supranational capacity. National governments (in the non-developed world) are paralysed through various tools and weapons of exploitation. The target of converting an independent country into a profitable market for organised forces of exploitation is achieved through coercive persuasion or use of force.

In the simplest way, globalisation is defined misleadingly as the integration of the national economy of a country with the global economy. Contrarily, however, the process of globalisation leads to assimilation of an autonomous economic community into a monopolistic and unified global economy controlled and regulated by the major powers of the world through the trinity of the World Bank (WB), the International Monetary Fund (IMF), and the World Trade Organisation (WTO). Thus, assuming that globalisation is denationalisation, deregulation of economic activity is blatantly wrong and grossly misleading. In fact, it is denationalisation of vital resources (natural, material and human) preserved in non-developed countries and reregulation of economic activities according to the dictates of major economic actors operating in the capitalist framework of maximisation of profit. Keeping in view the space occupied by multinational corporations in the fast-expanding global economy, globalisation is also termed as corporatisation.

It is important to understand that we are living in the age of mass deception where words have lost their true meaning, concepts have become blurred and scholars and intellectuals are increasingly losing their integrity. Major discourses and mega narratives have also sunk into a deep sea of ambiguities. The discourse on capitalism, globalisation and imperialism today has also become narrow and monopolistic. Chomsky, in the context of globalisation, observed:

> If we use the term neutrally, globalisation just means integration, welcome or not, depending on human consequences. In Western doctrinal systems, which prevail everywhere as a result of western power, the term has a somewhat different and narrower meaning. It refers to a specific form of international integration that has been pursued with particular intensity of private concentration of power and the interests of everyone else are incidental. With that terminology in place, the great mass of people around the world who object to these programmes can be labeled as adhering to 'anti-globalisation', as they always are. The force of the ideology and power is such that they even accept that ridiculous designation. They can then be identified as 'primitivists' who want to return to the 'stone age', to harm the poor, and with other terms of abuse with which we are familiar (37).

This is the reason why professional and public intellectuals, even in developing countries, are announcing that globalisation has come to stay and therefore, it should be welcomed. One is reminded of the famous lines of Marx: "The Philosophers have only interpreted the world in various ways; the point, however, is to change it" (*Thesis on Feuerbach*). These intellectuals or the *modern owls of Minerva* talk about the vast opportunities created by globalisation and ethics of the open competition unleashed by it. They argue that upcoming students and professionals should reap the fruits of globalisation by developing the spirit of competition and make efforts to outsmart fellow competitors. This is nothing but advocacy for inculcation and promotion of the spirit of Social Darwinism i.e., the ethics of survival of the fittest.

However, it is satisfying that there are organic intellectuals and scholars who do not share this pessimism that globalisation is irresistible. There are countries, communities and people in every nook and corner of the world (even in the most developed of countries) who are resisting the neoliberal capitalist onslaught disguised as globalisation. Anti-globalisation movements across the world are pointing to this fact. Oscar Wilde correctly observes: "One can give only an unbiased opinion about things that do not interest one, which is no doubt the reason an unbiased opinion is always valueless. The man who sees both sides of a question is a man who sees absolutely nothing" (Shakula 56). Bourgeoisie intellectuals may boast the opportunities created by globalisation for the educated and professional elite but the fact is that it has devastating effects on people in general and vulnerable segments of society in particular. This fact is echoed in various writings. For example; Mitra believes that "except for a very thin segment of population at the top, liberalisation is hurting everyone" (16).

It is important to mention that sometimes a myth or confusion is created by identifying globalisation with *Vasudeva Kutumbakam*[1]. In fact, it is gross injustice to our ancient civilization that contemporary globalisation is compared to this noble idea which calls for treating all human beings as members of one human family transcending political and ideological boundaries. Globalisation has an ideology and it cannot be seen in isolation from the basis of its origin i.e., capitalism which thrives on successive exploitation and usurpation of the natural rights of people. Capitalism passes through many stages and culminates in imperialism. Vladimir Lenin's book *Imperialism: The Highest Stage of Capitalism* (1916) has effectively exposed this fact. Perhaps the final culmination of capitalism is corporate globalisation which has marginalised the state to become subservient to market forces. This marginalisation of the state has led to peripheralisation of central concerns of the state for women, children and downtrodden sections of the society.

Globalisation and Structural Adjustments: Impact on Women

Globalisation, in its present form, necessitates structural adjustments by service-oriented public dispensations to provide space to profit-oriented economic forces. Bagchi observes: "We must recognise this fact that all phenomena associated with global interactions, such as trade, nationalism, state building and so on, have had contradictory influences on women's lives throughout history" (3). Globalisation, as superimposed and coercively enforced by the trinity of WB, IMF and WTO, compels sovereign states to open up their markets to global forces of production and trade. Thus, states have to opt for Structural Adjustment Programmes (SAP) supervised and regulated by the trinity. The SAP is characterised by

[1] *Vasudeva Kutumbakum* is a Sanskrit expression to denote that the whole world is one single family.

an *apparent* reduction in the role of the state in economic activity through deregulation and denationalisation. It entails unlimited freedom of import and export, lifting of tariffs, withdrawal of subsidies and heavy cutbacks in public expenditure, adoption of market-friendly fiscal exchange and credit policies, and criminalising resistance of indigenous people. The devastating effects of SAP are felt by the poor and the vulnerable in developing societies as is evident from what Dasgupta observes:

> The new instruments created by SAP for globalisation are diverse and encompass all aspects of women's lives in India... The traditional role of women in agriculture, livestock and animal husbandry, khadi and village industries including handicrafts, handlooms, fisheries, etc. is being undermined because mechanisation and automation are becoming prevalent in the market-based economy which will adversely affect the village-based traditional economy (3).

Ghosh finds:

> There is a mounting pressure to disband the public distribution system (PDS) and the welfare state, at the instance of the IMF and the World Bank, refuses to distribute food grains to millions of the poor at affordable prices. The system, aimed at benefitting people below the poverty line (BPL), seems to have no effect on the real poor. The off take from the PDS seems to have been deliberately slowed down due to irregular supplies, the bad quality of grains and the price hike, but this can never justify disbanding of the PDS; it actually demands a better administered PDS. He further observes that the cumulative effects of globalisation and the new market economy have resulted in an unprecedented hike in prices of staples like wheat and rice; for example, the retail prices of wheat and rice have increased by 12.5 per cent and 22 per cent respectively. Abolition of subsidies in the food and agricultural sector is likely to increase prices further and will invariably lead to the poor eating less and women the least (31).

It seems that an illusion of deregulation and decontrol is created; otherwise the state operates more fiercely in the post-globalisation phase. The frequent use of repressive apparatuses of the state in India and other countries of Asia, Africa and Latin America in the post-globalisation phase is a pointer to the fact.[2] The nature of the state as organiser of violence has

[2] In India, Nandigram, a place in West Bengal witnessed police violence against villagers opposing the land acquisition for SEZ. Interestingly, the farm lands were acquired by the then Left-front government of West Bengal which had been ruling over the state for more than three decades. The small farmers or land holders formed a committee called *Bhumi Uchhed Pratirodh Committee* (loosely translated

> not changed but its functions and responsibilities have undergone considerable change under SAP. In fact, the state has been converted into a private limited company. Under these circumstances it is difficult for the privatised or corporatised state to adopt poverty alleviation programmes, gender sensitive schemes, and affirmative action policies for downtrodden segments of its citizenry. Women and marginalised groups in all societies have to bear the brunt of this forced or deliberate withdrawal of the state. On the other hand, the state or the national government, controlled by the economic elite, exploits the situation to its advantage to evade its moral and social responsibilities towards the poor and the weak.

SAP is increasing global inequality and has given birth to the crisis of governance in the non-developed world. The deepening debt crisis, the issue of withdrawal by the state from public services like health and education etc. and growing social tensions are making the situation worse for the third world. Bacchus, in her insightful study, 'The effects of Globalisation on Women in Developing Nations' (3), has analysed the consequences of SAP in the following words:

> The world economic market depends on the flow of imports and exports between developed and developing nations. Throughout history, developing nations are faced with the lack of capital for the internal development of their country. Deficient funds are procured through loan from the World Bank and the IMF. In order to stabilise the flow of international capital, the World Bank and IMF enforce Structural Adjustment Loans (SALS) in developing nations. The amount of power and influence over the World Bank and IMF depends on the amount of capital being invested in the World Bank. Developed countries, like the USA and Japan, have an extremely large amounts of capital invested in the World Bank and are, therefore, major instruments in determining the actions and procedures taken by the IMF and World Bank. Thus the developed countries, by monopolising the decision-making in these bodies,

---

as Committee to Oppose Land Acquisition) and vehemently opposed the government's plan. Several villagers died in police firing and also in violence between the alleged CPI (M) cadres and Trinamool Congress. Finally the plan was cancelled by the West Bengal government and the left front rule came to an end in 2011 elections. Political commentators and analysts say that the Nandigram land acquisition program was mainly responsible for the defeat of the Left-front in the state. Similar incidents were reported from several other places in India, for example; Singur in West Bengal, Bhhatta Parsaull in Uttar Pradesh, and Jagatsinhgpur district of Orissa. All these places have witnessed opposition and protests taking violent shape and brutal use of force by the state. These kinds of anti-globalisation movements have become a universal phenomenon. The anti-globalisation movements are also termed as global justice movements.

enforce policies and programs on developing nations to trap them in an unending and perhaps never-payable debt crisis.

It is also found that the countries forced to surrender their right to self-determination in economic affairs are caught in a situation where they pay back more money to the World Bank and IMF in interest and repayment instalments than they receive from these neoliberal capitalist institutions. Arundhati Roy analyses the situation faced by India and observes: "we are forced to incur new debts in order to be able to repay our old ones. According to the *World Bank Annual Report (1998),* after the arithmetic, India paid The Bank $478 million more than it borrowed. Between 1993 and 1998, India paid The Bank $1.475 billion more than it received" (77).

Once the state is trapped under SAP, it becomes impossible for it to function as a welfare state committed to egalitarian principles. As a result, cutbacks in public expenditure become inevitable which adversely affects women and weaker sections. Sinha believes that globalisation hits the poor the hardest, and among them, it hits women even harder" (97). She looks at the impact of SAP on Indian women and points out, "Poverty alleviation programmes of the Government of India had included women as anganwadi workers and some 1.2 million workers were involved in rural areas providing support to women and children, but serious cutbacks have stopped the growth of this programme" (96).

Capitalism and Women

Globalisation should be viewed and analysed in the neoliberal capitalist paradigm of successive exploitation and what Harvey calls neoliberal design of 'accumulation through dispossession' (137-182). Thus globalisation, accompanied by liberalisation and privatisation, has created a large number of 'victims of globalisation'[3] in almost every nook and corner of the world. It is leading to global disparity, marginalisation or exclusion of indigenous people and communities. EM Wood agrees that globalisation based on openness and free trade has increased the interdependence of world's economies but he says that this openness and the so-called free trade are one sided. He observes, "Global capital

[3] Victims of globalisation includes small farmers, casual laborers, migrant workers, small shop-keepers, vendors, socially excluded groups, marginalised communities, internally displaced people like adivasis (forest communities of India), who are forced to leave their natural habitats where they have been living for thousands of years. Besides, all the affected people of ecological imbalance and environmental pollution are in fact the victims of globalisation.

actually benefits from the unevenness of the national economies, which allows it to exploit cheap labour and resources, while at the same time blocking competition from those low-cost economies" (18).

Capitalism and its supporting political manifestation i.e. liberalism or electoral democracy is nothing but a market-oriented ideology. It encourages and also creates conditions for consumerism, marketisation and commoditisation. The commoditisation of human labour is inherent in capitalism. Thus feminised commoditisation is the most visible manifestation of the contemporary capitalism/globalisation. The highly profitable and flourishing cosmetic business is involved in bluntly telling women everyday that they are ugly, dark, fat, heavy and not saleable in their present shapes. These features of the human body are often projected as problems and fairness creams, lotions, soaps, and beauty treatments are offered as readymade solutions to these problems. The psychological and emotional hurt generated by this capitalist ethic of business is most often ignored. This negativisation of body image or manipulated positivisation of beauty is a highly disturbing phenomenon unleashed by capitalist globalisation. This is also a noticeable fact that dark or black skin is projected as the symbol of ugliness and white skin as beauty personified. No one knows why black is considered the colour of mourning, sadness, evil and death and white the colour of peace, prosperity, innocence, perfection, honesty and good. Perhaps this manipulated consciousness about colours has a lot to do with imperialism and domination. One should keep in mind the white man's burden theory of colonialism and the contemporary standards of beauty set by the corporate world gives rise to a feeling of déjà vu.

The sole motto or objective of capitalism at all stages of history has been the maximisation of profit through excessive exploitation of resources – human and material. Therefore, looking for feminisation of progress and prosperity in the wake of globalisation is a futile exercise. However, the flag-bearers of globalisation often argue that it has led to empowerment of women through feminisation of labour. Nevertheless, this claim is not supported by reliable data, statistics or studies. And, in the absence of reliable gender-sensitive studies of globalisation, it becomes pretty easy for neoliberal intellectuals to mislead academia and people at large about the positive impact of globalisation on women.

Globalisation provides opportunity to the rich to get the best kind of everything in the market. For example, mineral water in a plastic bottle – packed and certified by the government that it is safe, fresh and rich in minerals. Thus the rich and the class that influences the decision-making process gets fresh water by paying more and more and on the other hand, the poor and the deprived, although more in number, are not able to influence decision-making and has to be content with contaminated water which sometimes contains arsenic and other dangerous substances. They are left to get ill and die. The masses, abandoned by the state, are left at the mercy of natural forces. However, it is also a fact that multinational companies (MNCs) earn a lot by selling medicines to these people.

Corporatisation and Women

Globalisation is so closely linked with privatisation that, at times, it becomes difficult to distinguish between the two. One fails to understand without dialectical analysis of the process of globalisation why there is so much emphasis on privatisation in globalisation? If globalisation is all about giving a free hand to MNCs to exploit, loot, plunder and maximise profit in the name of free trade, it is not globalisation but corporatisation.

MNCs or Transnational Corporations (TNCs) are the major beneficiaries and the most important players in the globalising economy. "By the beginning of the twenty-first century, TNCs accounted for two-thirds of world trade and now constitute half of the 100 largest economic entities in the world" (Greg, Hulme and Turner 164). The top 200 corporations' combined sales are bigger than the combined economies of all countries barring nine. That is, they surpass the combined economies of 182 countries. (Pratap n.pag). It is also interesting to note that multinational corporations, rather than creating jobs, have contributed to job cuts. According to a factsheet about multinationals, the top 200 corporate houses have been net job destroyers in recent years. Their combined global employment is 18.9 million – less than one-third of one per cent of the world's population. And not only are these corporate houses cutting workers, their CEOs are benefitting financially from these job cuts (Pratap n.pag).

These largest MNCs are based almost exclusively in advanced industrialised countries; ninety-nine of the 100 largest firms are based in the United States, Western Europe or Japan and more than five-sixth of all parent corporations are based in advanced industrial countries. John Agnew navigates into the geopolitics of globalisation and points out that 'globalisation has a home address.' He finds that globalisation has been deliberately fostered to perpetuate American hegemony. He challenges the tendency to neutralise globalisation's image by passing it off as an entirely technological, sociological or ideological phenomenon (127). He observes: "… the form that recent globalisation has taken is the result of political choices that can be reversed or redirected" (Agnew 127). Freeman cites the study conducted by Oloka-Onyango and Udagama (*Globalisation is Gendered*) and asserts, "Globalisation has drawn more women worldwide into the labour force, but women feature disproportionately among the most exploited workers. This has led to what has been called 'the feminisation of poverty". Women are under-represented in global economic decision-making bodies and over-represented among the victims of globalisation (155)."

Bagchi talks of gender discrimination that is intimately related to the unleashing of unbridled competition under globalisation. She argues:

> Men are seen as frontline fighters in economic warfare and women as only the sustaining camp followers. In a country like India, which was never gender-friendly, this has meant a rise in the relative degree of neglect of women's health and nutrition. With the advent of modern technology, it has become relatively easy and cheap to murder millions of female foetuses, and the adverse sex ratio in most states has gone down even further, particularly among girl children. It is not accidental that the supposedly dynamic economies of Gujarat and Haryana have taken the lead in the gruesome ritual of female foeticide (11).

For Mies 'housewifisation' of capitalist development comes in the wake of globalisation (110-115). She argues that the growing trend 'to invest in women' is nothing but an attempt to lower global labour costs through increasing the exploitability of women and instilling greater discipline into the labour force. There is one fundamental difference between capitalist exploitation of earlier days and in our time. Under globalisation, we find marginalisation rather than exploitation. This marginalisation is more visible in the countries of Asia, Africa and Latin America that are projected as the major beneficiaries of globalisation. However, it is pertinent to note that marginalisation is witnessed even in highly developed countries and cities. For example, Soja, in his insightful study of the socio-economic transformation of Los Angeles since the 1960s, found the growing poverty amid extravagant wealth within the city and referred to it as 'peripheralisation of the core' (Greig, Hulme, and Turner 166).

The myth of empowerment of women through feminisation of labour in MNCs is exploded when one sees the conditions of workers there. It seems that women working in MNCs do not come under the purview of International Labour Organisation (ILO) Conventions and other international and domestic legislation favoring workers in the globalising economy organised on the principles and prescriptions of neoliberal capitalist ethics. Thus, they do not get maternity leave, childcare leave and other attending benefits. In this situation, they have to delay pregnancy to avoid loss of their jobs in the wake of pregnancy (firing non-productive pregnant women workers and hiring new ones is a common practice in MNCs). Furthermore, achieving targets fixed by the employer and meeting deadlines creates a lot of pressure on women working in MNCs and other private companies. The horrific effects of these problems on women must be analysed by medical practitioners and researchers.

Outsourcing is a cardinal feature of modern globalising economy. It is often argued that outsourcing helps developing countries as it creates jobs, and brings investment and foreign capital. However, the facts relating to outsourcing present a highly disturbing scenario for the non-developed world. William Blum discusses the dark side of outsourcing as he points out that in December 1991, chief economist of the World Bank Lawrence Summers wrote an internal memo advising the bank to encourage the migration of 'the dirty industries' to less-developed countries because, among other reasons, health-impairing and death-causing pollution costs would be lower. He further wrote "I think, the economic logic behind dumping a load of toxic waste in the low-wage country is impeccable and we should face up to that (Blum 6)." In 1998, the US Vice President, Al Gore, is alleged to have put great pressure on South Africa, threatening trade sanctions if the government didn't cancel plans to use much cheaper generic AIDS drugs which would cut into US companies sales. With this kind of scenario, one can understand that globalisation will never benefit the poor, the marginalised or the vulnerable. Whatever benefits in the form of jobs and apparent prosperity poor countries and communities get is accidental and temporary. On the other hand, outsourcing and other tools of neoliberal capitalism can create environmental hazards, disturb biodiversity and impose the dangerous practice of monoculture. As a result, the poor and the weak, including women in the developing world, will be doubly victimised.

Privatisation and Women

Privatisation, in the words of Roy, is the transfer of productive public assets from the state to private companies. These assets include natural resources – earth, forest, water, air and so on. She emphasises that these are the assets that the state holds in trust for the people. She observes:

> In a country like India, seventy per cent of the population lives in rural areas. That's seven hundred million people. Their lives depend directly on access to natural resources. To snatch these away and sell them as stock to private companies is a process of barbaric dispossession on a scale that has no parallel in history (153).

Privatisation of jobs leads to lack of job security and remarkable increase in the number of low-paid and part-time jobs. The employer is benefitted by the competition among jobseekers who are ready to sell their labour at a low cost as they do not have any other means of subsistence. Under these circumstances, employers enjoy the benefit of surplus value and exploit the dependence and helplessness of workers. Thus, we find that what has been achieved after a long struggle is being lost in the wake of globalisation. Social security, working hours, minimum wage rules, participation of workers in the management of industry, etc. are flagrantly violated. Women workers in these situations are more vulnerable and prone to all forms of exploitation – physical, emotional and material.

Moreover, the Special Economic Zones (SEZs) have further compounded the problem of women workers. These Special Economic Zones have become prisons for labourers where they are held captive every day. The collective bargaining power of workers through their organised struggle under labour unions has been undermined. On the other hand, the bargaining or exploitative power of the capitalist economy has multiplied several folds since it is not possible for democratic dispensations to repeal labour laws and dismantle trade unions at one go. So-called egalitarian democracies of welfare states have found ways, via media, to give a free hand to foreign and other investors to exploit the cheap labour in these free trade zones. Sometimes, these SEZs become concentration camps for women workers trying to manage the pressure created by quota deadlines and production demands set by the employer on the one hand, and family responsibilities on the other.

The casualisation of labour under corporate globalisation is a dangerous trend. Like workers in the SEZs, casual laborers too are at high risk of all forms of exploitation. These casual laborers are employed by MNCs through subcontracting arrangements. The subcontractors sometimes project themselves as placement cells or manpower suppliers and exploit the victims of globalisation – migrant workers and women. On the other hand, MNCs or traders are benefitted by exploiting the vital asset of human capital without any liability.

Call centres in India have been projected as one of the most flourishing industries providing jobs to thousands. Roy perspicaciously analyses the most ignored and, often, the most invisible aspect of these call centres. She points out the humiliation of an ancient civilization by the corporate world in this *promising* industry. She talks about the call centre colleges where young English-speaking Indians are groomed to man backroom operations of giant transnational companies. In her own immutable words:

> On no account must the caller know that his or her enquiry is being attended to by an Indian sitting at a desk on the outskirts of Delhi. The call centre colleges train their students to speak with American and British accents. They have to read foreign papers so they can chitchat about the news or the weather. On duty, they have to change their given names. Sushma becomes Susie, Govind becomes Jerry, Advani becomes Andy. (Hi! I'm Andy. Gee, hot day, innit? Shoot, how can I help ya?) Actually its worse: Sushma becomes Mary. Govind becomes David. Perhaps Advani becomes Ulysses (182-183).

She reveals that the call centre workers are paid exactly one-tenth of the salaries of their counterparts abroad. She touches our sensitivity when she talks of this multi-million dollar industry describing it as "a multi-million dollar industry built on the bedrock of lies, false identities and racism" (Roy 182).

Globalisation has increased rural poverty which usually goes unnoticed. It was only after the horrifying news of mass suicides of farmers became public that the issue of rural poverty and deprivation got some space in the public discourse. The suicides were the result of the Structural Adjustment Programs imposed on India to open up the agricultural market, especially the seed sector to the MNCs. It changed the nature of cultivation in India as farm seeds were replaced by genetically engineered corporate seeds (sterilised to make them unusable for replanting). Moreover, these corporate seeds needed chemical fertilizers and pesticides sold and bought at heavy prices. Thus, it increased the cost of production and poor farmers were trapped in heavy debts which they were never able to pay. The state – the Sovereign Socialist Republic of India, now operating under the dictates of the neoliberal capitalist institutions, could not come to the rescue of the victims of its own policies and programs. The result was mass suicides of small farmers. This was the paradox or contradiction of globalisation or what Mojab calls "death and starvation in the midst of enormous wealth" (79).

Thanks to democratic compulsions, the Minister of State for Agriculture, Harish Rawat, had to acknowledge in the Rajya Sabha (the upper chamber of Indian Parliament) that a total of 2,90,740 farmers committed suicide during 1995–2011. This is data provided by the National Crime Record Bureau (NCRB) in its annual report on suicides, Accidental Deaths and Suicides (ADS) – 2011 in India (*The Indian Express*). It is, in fact, two farmers committing suicide every day. P Sainath of the national daily, *The Hindu*, observed: "We have been undergoing the largest catastrophe of our independent history – the suicide of nearly a quarter of a million farmers since 1995. We are talking of the largest recorded rate of suicides in human history" *(India Tribune).* Many newspapers published photos of women holding framed photographs of husbands who had been forced to commit suicide leaving unfortunate wives and children at the mercy of, perhaps, nobody. These sad and silent faces of these women spoke much louder about the impact of corporate globalisation on them.

Ghosh has rightly analysed the negative impact of globalisation on Indian agriculture as multinational corporations are fast finding routes to monopolise it. He reveals:

> Monsanto, a US-based company, has bought up several Indian seed companies like Mahyco, Parry Rallis, etc. Simultaneously, indigenous knowledge is also under threat and mustard, cotton and castor are to undergo genetic changes to be initiated later by Monsanto. Cargill, another US giant controlling 80 per cent of the world's agricultural trade, has already established a monopoly in the wheat trade and flour milling, leading to the closure of a large number of small scale flour-processing mills in the villages. One estimate shows that such a monopolistic onslaught will lead to the loss of livelihood of more than 100 million people, including a significant percentage of women (29)

Now, "developed countries like the US, whose hugely subsidised farm industry engages only two to three per cent of its total population, are using drop agricultural subsidies in order to make the market 'competitive'. Huge, mechanised corporate enterprises working thousands of acres of farmland want to compete with impoverished subsistence farmers who own only a couple of acres" (Roy 202). "In effect, India's rural economy is being garroted. Farmers who produce too much are in distress, farmers who produce too little are in distress and landless agricultural labour is out of work as big estates and farms lay off their workers. They are all flocking to the cities in search of employment" (Roy 29).

The negative impact of corporate globalisation on women is very difficult to understand in India partly because of the nature of exploitation involved in globalisation and partly because of the peculiar nature of the society. When a man loses job or livelihood in a rural area, he migrates to an urban area. If he goes with his family, the women and children have to live in pathetic locations and in an unsupportive environment. The competitive job market in cities causes mental and physical pressure on the migrant worker and the family. Women and children, especially, have to bear the brunt as they are subjected to domestic violence. Far from her village, in an alien city, women have to suffer silently. Unfortunately, we do not have authentic studies conducted from this perspective. The positive impact of globalisation on some women is highlighted but the silent sufferings of poor women do not become news.

Globalisation has led to urbanisation of jobs. Concentration of jobs in mega cities and lack of opportunities in rural areas causes migration of small farmers, landless laborers and other victims of the suicide economy of globalisation to urban centres. This phenomenon has increased at an alarming pace in the post-globalisation period. The mega cities or urban centres have their own capacities and limitations of accommodating these migrant workers. In the absence of clear guidelines and policies from the government and owing to democratic compulsions, civil authorities and local administration becomes helpless and we witness mushrooming of illegal colonies, slum areas and ghettos. The plethora of problems created by this phenomenon can be only indicated but cannot be dealt with here as it requires a detailed treatment with supportive data, statistics and reliable studies. However, it cannot be denied that this forced migration has led to rise in crime rate, traffic jams on roads, increasing rate of trafficking in women, domestic violence against women, serious health hazards and shrinking space for everyone leading to further social tensions and unrests.

Political philosopher Rousseau believes that culture, science, technology and property are the enemies of man. He observes: “Such was, or may well have been, the origin of society and law, which bound new fetters on the poor, and gave new powers to the rich; which irretrievably destroyed natural liberty, eternally fixed the law of property and inequality, converted clever usurpation into unalterable right, and, for the advantage of a few ambitious individuals, subjected all mankind to perpetual labour, slavery and wretchedness” (qtd. in Jain 221). The ethics of globalisation legitimise exploitation, convert clever usurpation of rights into a mantra for development and progress and, above all, criminalise poverty and delegitimise people’s movements for survival.

Globalisation is an ongoing process of consolidation of neoliberal capitalism through privatisation and corporatisation. It cannot be seen in isolation from the capitalist ethics of free trade and maximisation of profit. It has been made possible by bulldozing autonomous economic communities and their ethics of production and distribution. Its institutional backups and legal arrangements are strong enough to make states subservient to market forces. Privatisation of power, politics and economy is its inherent feature. There is no doubt that it creates wealth but it is not concerned with the just distribution of the wealth it generates. Its only concern is demolishing all barriers that come in the way of advancement of neoliberal capitalism. Marx has correctly said that capitalism develops its own grave digger. Globalisation is the final culmination of capitalism and one must be optimistic that it will disappear soon because of its own inherent contradictions. At the same time, it must be asserted that the gradual withdrawal of the state from its basic duties and responsibilities toward its citizens, especially women and other non-dominant segments of Indian society, must be protested through people's movements.

Works Cited

Agnew, John. "Globalisation Has a Home Address: The Geopolitics of Globalisation." *Globalisation's Contradictions: Geographies of Discipline, Destruction and Transformation.* Ed. Dennis Conway and Nik Heynen. London and New York: Routledge, 2006. Print.

Bacchus, Nazreen. "The Effects of Globalisation on Women in Developing Nations." Pace University, 2005. http//digitalcommons.pace.edu/honorscollege-theses/2.

Bagchi, Amiya Kumar. "Rich Men's Globalisation: How do Women and the Poor Fare?" *Perspectives in Women's Studies: Globalisation.* Ed. Malini Bhattacharya. New Delhi: Tulika Books, 2004. Print.

Blum, William. *Rouge State: A Guide to the World's Only Superpower.* London: Zed Books, 2003. Print.

Chomsky, Noam. "Where is the World Heading?" *Globalisation: One World Many Voices.* Ed. Samit Kar. New Delhi: Rawat Publications, n.d. Print.

Dasgupta, Kalpana. "Globalisation and Indian Women: Problems, Possibilities and Information Needs: An Overview." World Library and Information Congress: 69th IFLA General Conference and Council, 2013. http//webdoc.sub.gwdg.de/ebook/2003/vortraege/iv/ifla69/papers/600-dasgupta.pdf.13 Nov.2012.

Freeman, Michael. *Human Rights: An Interdisciplinary Approach.* Cambridge: Polity Press, 2003. Print.

Ghosh, Asish. "Women Agricultural Workers, Food Security and Globalisation." *Perspectives in Women's Studies: Globalisation.* Ed. Malini Bhattacharya. New Delhi: Tulika Books, 2004. Print.

Greig, Alastair, Hulme, David and Turner, Mark. *Challenging Global Inequality: Development Theory and Practice in the 21*st *Century.* New York: Palgrave Macmillian, 2007. Print.

Harvey, David. *The New Imperialism.* Oxford: University Press, 2003. Print.

India Tribune. 23 Jan. 2012. http://www.indiatribune.com.

Jain, M.P. *Political Theory.* New Delhi: Authors Guild Publications, 1993. Print.

Marx, Karl. *Thesis on Feuerbach.* Print.

Mies, M. "Gender and Global Capitalism." *Capitalism and Development.* Ed. L. Skliar. London: Routledge, 1994. Print.

Mitra, Ashok. "The World Organisation and the Status of Women." *Perspectives in Women's Studies: Globalisation.* Ed. Malini Bhattacharya. New Delhi: Tulika Books, 2004. Print.

Mojab, Shahrzad. "Gender, Political Islam and Imperialism." *The New Imperialists, Ideologies of Empire.* Ed. Coolin Mooers. Oxford: One World, n.d. Print.

Pratap, Ravindran. "The Power of MNCs over the Global Economy." *Business Line* 11 Sept. 2001. http//www.hidu.com/businessline/2001/09/11/Stories/041120nm.htm.15 Nov. 2012.

Roy, Arundhati. *The Algebra of Infinite Justice.* New Delhi: Penguin Books, 2002. Print.

Shukla, Rakesh. "Tiger and Terrorism." *Frontline* 14 Dec. 2002. Print.

Sinha, Indirani. "Women, Globalisation and Trafficking." *Perspectives in Women's Studies: Globalisation.* Ed. Malini Bhattacharya. New Delhi: Tulika Books, 2004. Print.

The Indian Express. New Delhi: 31 Aug. 2012. Print.

Wood, Ellen Meiksins. "Democracy as Ideology of Empire." *The New Imperialists: Ideologies of Empire.* Ed. Colin Moores. Oxford: One World, n.d. Print.

CHAPTER SIXTEEN

WOMEN EMPOWERMENT UNDER MICROFINANCE PROGRAMME: THEORY AND PRACTICE

DEBASHIS JODDAR

I. Introduction

In international development discourse, the most popular buzzwords have been microfinance, participation, poverty reduction and empowerment during last two decades. Moreover, these words have created a new dimension in the development policy dialogue of Third World Economies since the year 2000 with the announcement of Millennium Development Goals (MDGs) by United Nations Organisation (UNO).

In the literature, *economic development* has been defined as 'a process of improving the quality of all human lives' (Todaro 1994), emphasising three important aspects: Raising income and consumption, fostering self-esteem through institutions that promote human dignity and respect and increasing people's freedom. Nobel laureate Prof. Yunus (1998) is of the view that if the goals of economic development include improved standard of living, removal of poverty, access to dignified employment and reduction of inequality, it is quite natural to start with women as they constitute the majority of the poor, under-employed and economically and socially disadvantaged.

It is quite important to mention that in developing countries, women are the poorest of the poor as they experience hunger and poverty in much more intense ways than men. But traditional banks and financial institutions were gender biased and did not lend money to women. Microcredit as a tool of poverty alleviation and empowerment, has gained much credence in the theory of development. This programme targets women as clients and it is based on the premise that the generation of employment and income through micro-enterprises would empower women. The two core objectives of the policy are eradication of *poverty* through the provision of formal financial access under demand-based microcredit programmes by the formation of Self Help Groups (SHG) and *empowering* the group members by following the route of access of credit (*generation of income → increase in reliance → increase in decision-making power → increase in participation→ rise in empowerment level).*

There is no doubt that in the present scenario, women have gained much greater access to credit than the previous supply-led programme implying that the level of poverty can be reduced to some extent. But as far as the empowerment issue is concerned, the ultimate gain is not so easy, because, it is not only a multi-dimensional issue but also an interlinked ongoing process of change in power relations.

The term 'empowerment' includes economic, social and cultural as well as political aspects of marginalised people, especially women. Education, right over property, decision-making power, ownership, entitlement, etc. have been identified as the means of empowering women. In this light of development paradigm, all the above means are being translated as a set of human rights if we can conceptualise our mother, sister, wife, daughter, woman colleague, friend or any other female member of the society as human beings. However, the root of the problem lies in the mismatch in intra-household relations of consumption, production and distribution of the household economy. Every household consumes goods and services both from within and outside the household but the decision-making power of family members related to purchasing of goods as well as ratio of consumption are determined by some traditional norms that are obviously gender biased. Similarly, households produce goods and services for consumption and for selling. There is another type of production – the production of human beings. It entails biological as well as social reproduction and physical and economic maintenance. The intra-household production relation is also an example of traditional socio-cultural norms in which women are considered inferior to men. The household also distributes time and resources according to consumption and production rules among household members in specific fields and in terms of hours of work. These distributive norms are set by ideology, lineage systems, kinship systems, religious codes, age, marital status, etc. which are invariably gender biased (Barua 2005).

In less developed countries, the most deep-rooted form of inequality is reflected through caste and gender relations. Gender relations are determined by interplay of factors in which societal structures, historical specificity, cultural norms and practices, political ideologies and economic conditions prevalent in a society play a dominant role. A patriarchal structure of society is characterised by male domination and female subordination. Social conditioning of women ensures that they remain inferior to men in different spheres of decision-making. One of the grossest forms of gender discrimination is found in various sexual attitudes, practices and forms of exploitation. Norms of sexual behavior are much more stringent in case of women. Increasing commercialisation and growing consumerism have opened newer avenues for sexual exploitation of women. Like all other commodities, women have also been identified as a 'commodity'. The advertising industry and sales promotion mechanisms fully exploit the bodily/physical charm of women.

In this paper, an attempt has been made to analyse the relation between microfinance and women's empowerment through four subsections. Sections II and III deal with theoretical ideas and relevance of microfinance respectively. Empirical evidence has been discussed in section IV and finally, concluding remarks in section V.

II. Theoretical Idea

Microfinance

Right from the 1950s, it has been the effort of governments in developing countries, donor agencies and international funding organisations to break the vicious cycle of poverty by providing subsidised credit to poor people. This old paradigm of microcredit programmes was usually accompanied by high loan defaults, high losses and general inability to reach the poor. In the history of microfinance, the 1980s represented a turning point with the shift in the paradigm from old to new demand-based programmes (Robinson 2001). The new microfinance operates on the basis of the principle 'Borrower knows best'. Under the new microfinance scheme, the operational strategy involves several features such as simple procedures for reviewing and approving loan applications, delivery of credit and credit-related services in a convenient and user-friendly way at commercial rates of interest, quick disbursement of small and short-term loans, clear loan repayment procedures, maintenance of high repayment rate, incentives like access to large loans immediately following successful repayment of first loan, etc. (Microcredit Summit Draft Declaration, 1997).

Microcredit has been defined by the Microcredit Summit (1997) as "programmes that provide credit for self-employment and other finance and business services (including savings and technical assistance) to very poor persons" (Draft Declaration and Plan of Action). Microcredit initially focused only on provision of loans to the very poor. The push to microfinance came with the recognition that households can benefit from accesses to financial services. The term was more broadly defined (at first the focus was mainly on savings) and included services other than credit for microenterprises. While the words microcredit and microfinance are often used interchangeably, they have different connotations. The word 'microcredit' refers to small loans whereas the broader term 'microfinance' signifies efforts to collect savings from low-income households, to provide insurance (micro-insurance), and also to help in distributing and marketing clients' output. With the change in language from microcredit to microfinance, there has been a change in orientation, towards 'less poor' households and towards establishment of commercially oriented, fully regulated financial entities (Armendariz and Morduch 2005).

The task force on Supportive Policy and Regulatory Framework for Microfinance has defined Microfinance as the 'Provision of thrift, credit and other financial services and products of very small amounts to the poor in rural, semi-urban or urban areas for enabling them to raise their income levels and improve living standards'. The characteristics of microfinance programmes include:

- Small amounts of loans and savings
- Short loan terms (usually up to one year)
- Payment schedules featuring frequent installments (or frequent deposits)
- Installments made up of both interest and principal
- High interest rates on credit (higher than commercial bank rates but lower than loan shark rates) reflecting the labour-intensive work associated with making small loans and allowing the microfinance intermediary to become sustainable over time
- Easy access to the microfinance intermediary, saving the client time and money, while also permitting the intermediary to better know the client in his/her home/business context
- Simple application forms which are easy to complete
- Short processing periods (between the completion of the application and the disbursement of the loan)
- sThe availability of repeat loans in higher amounts for clients who pay on time
- The use of tapered interest rates (decreasing interest rates over several loan cycles) as an incentive to repay on time. As larger size loans are less costly to the MFI, some lenders charge higher interest rates on small credit amounts and lower ones on larger credits

- No collateral is required contrary to formal banking practices. To replace collateral (which low-income people generally do not have), microfinance intermediaries use alternative method, such as assessment of clients' repayment potential by running cash flow analyses based on the stream of cash flow generated by the activities for which loans are taken; all enterprise/household income and expenditure items; individual or group guarantees (solidarity groups) and compulsory savings schemes (Murray and Boros 2002).

Now we shall try to explore the ideas of two great Indian thinkers – Tagore and Gandhiji. It has been found that the idea of microcredit on the basis of Self Help Groups is not a new concept. It was found that as far back as 1894, Tagore started microcredit programmes to finance rural development activities on cooperative schemes with the objective of freeing really poor farmers from the grip of money lenders. Tagore stresses, "The most urgent need in our country is not to place begging bowls in their hands but to make them confident of their own power; to make them realise that a man united with others is a complete entity whereas an alienated individual is but a powerless fragment." (Tagore 313). This is also the notion microfinance.

According to Gandhiji, the term '*swadeshi*' is translated as 'self-reliance' referring to material self-sufficiency as much as to self-esteem of the human being. The other term '*swaraj*' signifies 'self-rule'. Finally, there exists no difference between the Gandhian spirit of self-determination and self-management and the present notion of SHG. Both ideas imply that personal improvement is a process that is determined by relationships of solidarity and cooperation. (Guerin and Palier 2005)

Self Help Groups (SHGs)

Since 1970, worldwide, policymakers and academicians started thinking about how to link development programmes to poor women. There has been considerable rethinking on the impact and potential of self help groups (SHGs) on rural poverty and empowerment of women since the Grameen Bank first pioneered the concept in 1979. Professor Mohammad Yunus, Nobel laureate in peace, came up with the idea of providing small loans to poor households in the neighbourhood, particularly poor women households – those unable to provide collateral. The problem of women's access to credit was given emphasis at the first International Women's Conference in Mexico in 1975, which resulted in the establishment of the Women's World Banking network. In 1985, during the second International Women's Conference in Nairobi, there was a mushrooming of government and non-government income generating programmes for women most of which included savings and credit.

The microfinance methodologies can be classified into five groups (Satish 2005):

(i) *Grameen and Solidarity model*: People form groups of three to eight persons subject to the condition that each of them would take responsibility for the lending and other financial operations for the other members of the group – the celebrated 'joint liability clause'. Lending and repayment – both are directly related to individuals but with the guarantee of the other group members. This methodology has been adopted by Grameen Bank in Bangladesh, BoncoSol of Bolivia and most of the solidarity group models in Latin America.

(ii) *The Group Approach*: It delegates the entire financial process to the group rather than the financial institutions. The group considers the issue of savings, loans and repayment of loans. These groups are, in turn, linked to a financial or microfinance institution to receive additional funds and to deposit their savings. Self help groups' bank linkage programme in India, the PHBK project in Indonesia and the Chikola groups of K-REP in Kenya are examples.

(iii) Individual Credit: This methodology is predominant in the BRI-Unit Desa in Indonesia and priority sector lending by banks in India. According to this methodology, loan appraisal, loan disbursements, loan repayments as well as savings collection are done on an individual basis.

(iv) Community banking (village banks): This model is an extension of the group approach where the basic financial needs are met through the community banking system. These banks borrow from the programme-implementing institution for lending to the members. The most important example is the Village Bank of Finca in Latin America which had been replicated in Africa and central Asia.

(v) Credit Unions and Cooperatives: These are member-owned organisations that provide credit and other financial services to their members. SANSA of Sri Lanka is a successful example of a rural credit cooperative providing microfinance services.

Empowerment

Empowerment of women, a Western concept, emerges as an outcome of several important critiques and debates generated by the women's movement throughout the world and especially by Third World feminists. As an articulated goal of development, it has been gaining popularity over the last few decades. Its source can be traced to the interaction between femininity and the concept of popular education as developed in Latin America in the 1970s.

'Empowerment of women', in its simplest form, means the manifestation of redistribution of power that challenges patriarchal ideology and male dominance. It is a process that enables women to gain access to and control material as well as information resources. On one hand, it is a process and on the other hand, it is the result of the process. It is reflected through the transformation of the structures or institutions that reinforce and perpetuate gender discrimination. In practical terms, empowerment implies the process of challenging of existing inequality and power relations and of gaining greater control over sources of power by the underprivileged (De 2009).

Random House dictionary states that the term 'empowerment' originates from the term empower which implies 'to give power or authority'. The key elements are 'enabling' and 'providing power'. Power can take various forms – 'power to', 'power with', 'power from within' and 'power over' view. In social sciences, power over view takes the dominant position.

According to Foucault (1982), power relations are a multiple which is constituted in a system of social network. Such a view of power relations cannot be altered without bringing about major structural change. Peggy (1989) looked at empowerment from the power angle and made a distinction between 'personal power' (power for) and 'role power' (power over). She conceives empowerment as 'a spectrum of political activity ranging from acts of individual resistance to mass political mobilisations that challenge the basic power relations in our society' (cited in Panda 4). Batliwala (1994) also defines empowerment as the process of challenging existing power relations and of gaining greater control over the sources of power.

Conger and Kanungo, (1988), distinguishes empowerment at the organisational level as well as interpersonal level. At the interpersonal level, power of an individual/group lies in its official position, personal characteristics, expertise and opportunity to access specified knowledge. At the organisational level, the primary source of an individual/group's power over an organisation is the individual/group's strength of resources that is valued by the organisation.

For women in India, empowerment exists in several realms: Personal, familial, economic and political. The theoretical framework to study the women's empowerment in the domestic sphere makes two key assumptions: (i) Empowerment is basically a property of social or cultural systems rather than of individual experiences and traits (Smith, 1989). (ii)

Empowerment is multi-dimensional, with imperfect associations between its different dimensions. The extent of women's power is often dissimilar in the private and public sphere as well as in the social, economic, political and psychological sectors. For example, although women are often found to be economically powerful, as they are engaged in income-generating activities, they are often sexually and socially submissive to their husbands and peripheral to the political process (Milne, 1982; Knodal, Chamratrilhirang and Debsvalya, 1987; Mueke, 1992). In this respect, it should be mentioned that Karl (1995) argues that political participation is central to the empowerment process. She emphasises the four inter-related and mutually reinforcing components of empowerment – collective awareness building, capacity building and skill development, participation and decision-making power and actions to bring gender equality. There are three contrasting paradigms – financial self-sustainability paradigm, poverty alleviation paradigm and feminist empowerment paradigm – related to microfinance and gender (Mayoux 5-9).

In the first paradigm, empowerment is defined as economic empowerment, expansion of individual choices and capacities for self-reliance and the second paradigm states that empowerment implies increased well-being, community development and self-sufficiency. The second paradigm finds financial and other services targeted at women for fulfilling their needs for income and employment which would automatically lead the way towards women's empowerment by reducing gender inequality. In the feminist paradigm, empowerment goes beyond economic betterment and well-being and indicates transformation of power relations throughout society. Mayoux suggests that empowerment is a process of internal change. While analysing women's empowerment, she gives a clear idea of different forms of power.

'Power to' implies increased access to income and productive assets, increased mobility, improvement of health and nutrition, meaning, thereby, increased capacity for change of individual women.

'Power over' indicates reduction in obstacles to change at household and community levels which is possible through control over use of loan, productive assets and household property, ability to defend self against violence in household and community, etc. *'Power with'* refers to the increase in solidarity with other women for change at household, community and macro levels. This form of power – 'power with' – can be gained through increase in network, participation in movements to challenge gender subordination and increase in expenditure for children, especially for girl children (Mayoux 45).

'Power within' refers to increased will for change of individual women i.e. increase in confidence, autonomy and willingness to take decision and change in consciousness to challenge gender subordination (Mayoux 45).

Then we have to move to the concept of 'spaces', which allow a person (woman) to do what she actually intends to do in place/freedom/margin. Initially, the allotment of spaces is determined by micro (domestic) and macro (social) environment within which she lives. Finally, different dimensions of these two basic elements – physical, biological, psychological, economic, socio-cultural and political – establish a person's ability to act or work, her behavioral pattern, state of reactiveness, level of humanity and also the eventuality of being a dignified human character both within and outside the family. (Deshmukh-Ranadive 2005)

Nobel laureate Professor Amartya Sen (1981), in his famous book, *Poverty and Famine*, describes the notion of empowerment in terms of entitlement i.e. right to equitable share of resources for men and women. But many recent definitions of empowerment adopted Prof Sen's (1999) 'development as freedom' approach as a starting point of empowerment. According to him, the goal of economic development must be to increase choices rather than to achieve a certain set of indicators. He (1993) also explains that the freedom to lead different types of lives is reflected in the personal capability set, which in turn depends on personal characteristics and social arrangements along with other factors.

Kabeer (1999) uses this approach in her definition of empowerment as she conceives 'power' in terms of 'ability to make choices' (436). She opines that empowerment is a process of change by which those who have been denied the ability to make choices acquire such ability. In Kabeer's language of empowerment, the notion of 'choice' has been qualified in a number of ways (437). First, choice necessarily implies the possibility of alternatives. The second set of qualifications referred to the consequences of choice, the need to distinguish between strategic life choices and second-order choices. Kabeer emphasises strategic life choices which are critical for people to live as they want (such as choice of livelihood, whether and who to marry, whether to have children etc.). According to her, empowerment refers to the expansion of people's ability to make strategic life choices in a context where this ability was denied to them previously. Kabeer also views the ability to exercise choice in terms of three inter-related dimensions – resource (pre-condition), agency (processes of decision-making as well as less measurable manifestation of agency such as negotiation, deception and manipulation) and achievements (well-being outcomes) (437-438). Resources include access as well as future claims to material, human and social resources including social relationships (family, market and community). Agency means the ability to define one's goals and act upon them and it also encompasses the meaning, motivation and purpose which individuals bring to their activity or their power 'within. Resources and agency constitute what Sen (1985) refers to as capabilities: The potential people have for living the lives they want, of achieving valued ways of 'being and doing'. Sen's idea of functioning refers to all possible ways of 'being and doing' which are valued by people in a given context, and 'functioning achievements' implies the particular ways of being and doing which are realised by different individuals. It is generally argued that proper nourishment, good health, adequate shelters constitutes primary functioning and tend to be universally valued. If there exists systematic gender differences in these very basic functioning achievements, they can be regarded as evidence of inequalities in underlying capabilities rather than differences in preferences. This is the strategy adopted by Sen (1990).

In a complementary approach, Chen and Mahmud (1995) highlights different dimensions of empowerment through four types of changes – *material, cognitive, perceptual* and *relational. Material empowerment* occurs through the expansion of the resource base of women. *Cognitive empowerment* results from women's recognition of their own abilities and skills, indicated by greater self-esteem and self-confidence. *Perceptual empowerment* takes place through changes in how others perceive them, indicated by increased social prestige and value. *Relational empowerment* occurs through changes in gender relations within the family and in the broader society, indicated by reduction in gender inequality in relationships. They identify empowerment as a process of external change, namely, through the changed perceptions of others and through changed gender relations, whereas Kabeer emphasises that internal changes are important for empowerment (cited in Mahmud 587).

World Bank (2005) defines empowerment as the expansion of assets and capabilities of poor people to participate in, negotiate with, influence, control and hold accountable institutions that affect their lives (5). There are four elements of empowerment – access to information, inclusion and participation, accountability and local organisational capacity – that underlie institutional reforms. Four aspects of this conceptual framework need to be highlighted. First, empowerment is fundamentally a relational concept that emerges through the rights, rules, resources, incentives norms, behaviour and processes governing the interactions between poor people and their environment in different arenas (state, community and household levels). Secondly, poor people's assets and capabilities are critical in helping them break through the constraints of powerlessness and voicelessness. Thirdly, empowerment of poor people on a large scale requires both top-down and bottom-up changes in organisational processes. Finally, interventions vary depending on the nature of constraints and barriers, on what is feasible, and on the development outcomes desired.

A review of these definitions suggests that empowerment is a process which has certain key elements: Power, autonomy and self-reliance, entitlement, participation and process of building awareness and capacity. All these elements are applicable at the individual and group levels to understand the process of empowerment (Panda 7).

III. Relevance

In terms of the Indian context, women in the rural industrial setup were not able to actively participate in income-generating economic activities owing to historical and socio-cultural reasons. Illiteracy, low level of skills, lack of access to training and credit facilities coupled with lack of spirit of entrepreneurship, invisible contribution to family economy, restricted mobility etc. as a result of gender bias are some of the contributing factors for it (Puhazhendi & Jayaraman 1999). This has induced many non-Government organisations (NGOs) to enter the rural credit scene by organising the poor, especially women in rural areas into informal groups for self help through mutual aid since the early eighties. Almost at the same time (1982-83), the group approach was adopted by the Government under a programme called Development of Women and Children in Rural Areas (DWCRA), an exclusive programme for women as a sub-component of Integrated Rural Development Programme (IRDP) with a view to using the collective strength of groups of poor women to break social barriers that restricted their income-generating and self-fulfilling opportunities. NABARD initiated projects on SHGs as a channel for delivery of microfinance in the late 1980s. SHGs got a distinctive identity 1992 onwards due to the leadership role provided by the Reserve Bank of India (RBI) and NABARD. It would also lead to women's empowerment since microfinance programmes have mostly targeted women as clients (Littlefield, Morduch and Hashemi 2003; Cheston and Kuhn 2002).

The microfinance movement, introduced in India with the introduction of the pilot project of NABARD in 1992-93 for linking Self Help Groups (SHGs) with credit institutions, hoped to evolve supplementary credit strategies for reaching the rural poor in India. With the IRDP being subsumed in the Swarnajayanti Gram Swarojgar Yojana (SGSY), effective from 1st April, 1999, the formation of SHGs has become the principal mode of poverty alleviation through self help and development of micro-enterprise.

The acceptance of gender equality in the Constitution of independent India provided women with a basis for a new identity as full citizens of the republic and a source of their rights to equality, dignity and justice in other spheres of life (Gupta and Chattopadhyay 111). Since the inception of the Fifth Five Year Plan (1974-79) in India, women's interests have been highlighted in national policy (WCD 2001).The SHG model was introduced as the core strategy to achieve empowerment in the Ninth Plan (1997-2002) with the objective to organise women into SHG and thus mark the beginning of a major process of empowering women (Planning Commission 1997).

In India, the primary initiative to organise SHGs was undertaken by NGOs in the early 1980s. But the first official interest in informal group lending took shape during 1986-87 on the initiative of NABARD. As a part of this broad mandate, NABARD initiated projects on SHGs as a channel for delivery of microfinance in the late 1980s. SHGs got a distinctive identity 1992 onwards due to the leadership role provided by the Reserve Bank of India (RBI) and NABARD. In 1996, NABARD launched a nationwide pilot project to link the SHGs to the bank. In 1998, NABARD tied up with GTZ (German Agency for Technical Collaboration) Rural Finance Programme to support the self help groups' Bank Linkage Programme (SBLP). Most importantly, with IRDP being subsumed in SGSY (Swarnajayanti Gram Swarojgar Yojana) effective on and from 1.4.1999, formation of self help groups and development of micro-enterprise has become the principal mode of poverty alleviation by the Government of India.

As far as West Bengal economy is concerned, the microfinance movement was initiated later compared to southern states but has gained momentum quickly. This movement is becoming increasingly popular as a way to mobilise poor communities through the provision of loans based on group solidarity instead of formal collateral. Both bank-led and NGO-led self help groups have been working in the state for almost two decades. From 2007-08 to 2011-12, the number of deposit-linked SHGs increased from 5,22,201 to 6,85,448 i.e. 1.31 times while that of credit-linked SHGs increased at a marginally higher rate from 52,558 to 99,379 i.e. 1.89 times. During this period, the number of credit-linked SHGs out of the total number of savings-linked SHGs increased from 10.06 to 14.5 percent. Cumulative amount of credit increased appreciably from Rs.22535.64 lakh to Rs. 55136.55 lakh during this period whereas amount of savings of

SHGs decreased from 46549.19 lakh to 37694.40 lakh. This indicates that credit-deposit ratio increased from 0.48 to 1.46 (NABARD 2008 and 2012).

In West Bengal, members of SHGs promoted by different agencies like NGOs, banks, co-operatives, panchayets etc. almost always belong to the female category. In DWCRA SHG linkage programmes, all members of SHGs are usually women. About 90 per cent SHGs consist exclusively of women in SHG-bank linkage programmes in West Bengal. In SHG-NGO linkage programmes, more than 90 percent SHGs consist of women. This is also true of SHG-PACS linkage projects. As regards employment, SHGs of female category in West Bengal almost always consist of self-employed women working in traditional village and cottage industries which are agro-based like khadi, handloom, handicraft, sericulture, coir, wool or silk, spinning, leather and leather products, tailoring, agarbati and candle making and other rural micro industries related to the processing of cereals and pulses, *gur*, molasses, processing of fruits, animal husbandry, etc.

IV. Empirical Evidence

The study is based on data obtained from field surveys (primary source). Final field survey has been conducted on 600 households. The sample consists of one core group and one control group, each having 300 households. The survey has been conducted in six districts of West Bengal – Howrah, Hooghly, North 24 Paraganas, Birbhum, Coochbehar and Nadia based on the concentration of SHGs. It was conducted between September, 2012 to February, 2013.

In this section, an attempt has been made to study empirically women's empowerment – a participatory multidimensional process – through the *generation of income, ownership of assets, decision-making power, control over credit* and *awareness and participation,* whereby women become able to organise themselves to increase their economic self-reliance and to assert their independent rights on assets and control over resources, which will assist in challenging and eliminating their subordination – based on their participatory role in microcredit programmes through decentralised channels (SHGs) promoted by various agencies in the context of West Bengal.

Every qualitative element of empowerment has been measured on the five-point Likert scale: Very high (5), High (4), Medium (3), Fair (2) and Low (1). All elements of empowerment are assumed to have equal weight. The mean (AM) of all elements' levels indicate the empowerment value of the selected women. The highest and lowest level of empowerment should be 5 and 1 respectively. Members having level of empowerment 4 and above are termed 'highly empowered' women; those having scores between 2.5 and 3.99 are referred to as 'empowered' and members securing values less than the median score i.e. below 2.5 (50% of 5 point scale) are regarded as 'disempowered' persons.

The results (Table-1) indicate that the average level of empowerment of the core group varies from 2.47 to 4.83 whereas that of control group varies from 1.54 to 2.41. The data also reveals that 31% members of the core group have high level of empowerment, 53% belong to medium category of empowerment and the remaining 16% members of the core group are considered disempowered women. On the other hand, the percentage of 'highly empowered' and 'empowered' women of the control group together constitute only about 30% of all group members. This shows that 70% women who have not yet participated in microfinance programmes fall in the category of disempowered women.

Table-1. Level of Empowerment

Type of Sample	Highest Value of Empowerment	Lowest Value of Empowerment	% of People Having Level of Empowerment		
			High	Medium	Low
Core Group	4.83	2.47	31	53	16
Control Group	2.41	1.54		30	70

Source: Primary survey

While calculating the average score of each element of empowerment for 300 members of the core group, it has been observed (Table-2) that the average score of two elements, namely, 'ability to generate income from microcredit' and 'control over income' is more than 4. Taking into consideration the elements, 'ownership of assets' and 'decision-making power', the study finds that the average score of the elements of empowerment ranges between 3.5 and 3.99. The remarkable feature of the study is that the average scores of the elements, 'level of participation' and 'level of awareness', are relatively low. This demonstrates that although microcredit has widened the scope for increasing income and control over loan substantially, members are still lagging behind as far as social and political awareness and participation in the same sphere are concerned.

Table-2: Average Values of Elements of Empowerment of Core Group

Ability to generate income from microcredit	Ownership of assets	Decision-making power	Control over income	Level of awareness	Level of participation
4.24	3.65	3.79	4.22	3.01	2.67

Source: Primary survey

V: Conclusion

An attempt to link microfinance and women's empowerment has been made through theoretical as well as empirical discussion. But after this long journey, it is understood that the issue of empowerment is a complex issue to track and trace. It changes dimensions with respect to economic, social, cultural and political situations. There is no doubt that with the intervention of microfinance programmes, the standard of living of households has become better. Microfinance can also improve long-term development as women are the main brokers of children's health, nutrition and education. In particular, with increasing scope for income generation, women can access asset ownership as well as decision-making power and obvious entitlements in a sustained manner. This, in turn, should act against domestic violence and as an instrument for women to promote their rights, improve their bargaining power and widen spaces at the individual level in the domestic arena. But the question is to what extent microfinance can contribute to changes in legal rights, socio-cultural norms and private and political spaces. These issues should be revisited regularly until concrete and honest effort is made by concerned institutional or organisational authorities to incorporate them into programmes. However, unless a change in these spaces widens mental spaces of women and men in our society, one cannot say that the process of empowerment has been set in motion. It should be remembered that this is not a zero sum game where one person incurs loss for giving to another. In this context, we should remember Tagore's popular novel *Ghare-Baire*, where Nikhilesh has given enough space to his wife Bimala so that she can express herself with self-consciousness, passion, self-ideology and self-instinct. He has also given her complete freedom of choice that can make her a complete entity. The novel gives a clue to the world and helps solve the present puzzle. Nikhilesh eventually gives up his life in order to make Bimala's life more spacious.

Works Cited

Armendariz, B. and J.Morduch. *The Economics of Microfinance*. New Delhi: PHI Learning Private Limited, 2005. Print.

Barua, Neera., Joy Deshmukh-Ranadive and Ranjani K. Murthy, eds. *Micro-Credit, Poverty and Empowerment Linking The Triad.* New Delhi: Sage Publications India Pvt Ltd, 2005. Print.

Batliwala, S. "The Meaning of Women's Empowerment: New Concepts from Action". *Population Policies Reconsidered: Health, Empowerment and Rights*. Ed. Gita Sen et.al. Harvard Center for Population and Development Studies, 1994. Print.

Chen, M. and S. Mahmud. "Assessing Change in Women's Lives: A Conceptual Framework". Working Paper No. 2, BRAC-ICDDRB, Joint Research Project. Dhaka, 1995. Print.

Cheston, S. and L.Kuhn (2002): "Empowering women through microfinance".http://www.opportunity.org/files/Empowering%20Women%20Paper%20FINAL pdf UNIFEM.13 Feb.2004.Web.5 Jan.2009.

Conger, J. A. and R. N. Kanungo. "The Empowerment Process: Integrating Theory and Practice." *Academy of Management Review.*13.3(1988): 471-482. Print.

De, Sudipta. *Self-help Group Participation and Women's Empowerment: A Study on West Bengal.* Kolkata: Unpublished Thesis (Submitted to the University of Calcutta), 2009. Print.

Department of Women and Child Development. *National Policy for the empowerment of Women (2001)*. New Delhi: WCD.Web. http://wcd.nic.in. 15 May, 2005.

Deshmukh Ranadive, Joy. "Can Microfinance Empower Women?" *Microfinance Challenges: Empowerment or Disempowerment of the Poor?*. Ed. Isabelle Guerin. and Jane Palier. Pondicherry: French Institute Pondicherry, 2005. http//www.ifpindia.org. Web. 12 August, 2011.

Foucault, M. "The Subject of Power". *Michael Foucault: Beyond Structuralism*. Ed. Dreyfus and P. Rabinow. Brighton: Harvester P, 1982. Print.

Gupta, M.and R. Chattopadhyay. "Law and Violence against Women". *The Changing Status of Women in West Bengal 1970-2000.* Ed. Jashodhara Bagchi. New Delh: Sage publications, 2004. Print.

Guerin, Isabelle and Jane Palier. eds. "General Introduction". *Microfinance Challenges: Empowerment or Disempowerment of the Poor?* Pondicherry: French Institute, Pondicherry, 005. http//www. ifpindia.org. Web .12 August, 011.

Kabeer, N. "Resources, Agency, Achievements: Reflections on the Measurement of Women's Empowerment". *Development and Change*. 30 (1999): 435-464. Print.

Karl, M. *Women and Empowerment: Participation and Decision Making.* London: Zed Books, 1995. Print.

Knodal, J.A. Chamratrilhirang and N.Debsvalya.(1987): "Thailand's Reproductive Revolution: Rapid Fertiity Decline in a Third-World Setting". Madison: University of Wisconsin Press. *Measuring Empowerment: Cross Disciplinary Perspective.* Ed. Deepa Narayan. Washington DC: World Bank, 1987. Print.

Littlefield, E., J. Morduch and S. Hashemi. "Is microfinance an effective strategy to reach the millennium development goals?"*CGAP*. Focus Note Series, No 24 (2003). Print.

Mahmud, S. "Actually how Empowering is Microcredit?" *Development and Change*. 34.4(2003): 577-605. Print.

Mayoux, L. "Road to the Foot of the Mountain – But Reaching for the Sun". *PALS Adventures and Challenges*. November, 2005. Print.

Milne, C. (1982): *Asanti Market Women*. Video Recording, Granada Television International.http://www.therai.org.uk/film/catalogue 2/11_ ashante.html.

Measuring Empowerment: Cross Disciplinary Perspective. Ed. Deepa Narayan. Washington DC: World Bank, 1987. Print.

Mueke, M.A. "Mother Sold Food, Daughter Sells Her Body: The Cultural Contuity of Prostitution". *Social Science and Medicine*. 35.7(1992): 891- 901. *Measuring Empowerment: Cross Disciplinary Perspective.* Ed. Deepa Narayan. Washington DC: World Bank, 1987. Print.

Murray, U. and R. Boros. *A Guide to Gender Sensitive Microfinance.* Rome:Food and Agricultural Organisation of the United Nations (FAO), 2002. Print.

NABARD. Status of Microfinance in India 2007-08. Mumbai: *http://www.nabard.org.* Web. 13 July. 2010.

—. Status of Microfinance in India 2011-12. Mumbai: *http://www.nabard.org.* Web. 9 Jan. 2013.

Panda, S.M. (2000): "Women's Empowerment Through NGO Interventions: A Framework for Assessment". Working Paper No. 145, Institute of Rural Management, Anand, 2000. 4. Print.

Peggy, A. "The Empowerment of Women". *The Women and International Development Annual*. Ed. R. Gallin, M. Aronoff and A. Ferguson. Boulder: Westview Press, 1989. Print.

Puhazhendi, V. and D. Jayaraman. "Increasing Women's Participation and Employment Generation among Rural Poor: An Approach through

Informal Groups". *Indian Journal of Agricultural Economics,* 54.3(July-September 1999). Print.

Planning Commission. *Ninth Five Year Plan 1997-2002*. Web. New Delhi Government of India, *http://planningcommission.nic.in/plans/planrel/fiveyr/welcome.html.* 20 June, 2005.

Robinson, M.S. *Microfinance Revolution: Sustainable Finance for the Poor*. Washington: World Bank, 2001.

Satish, P. "Mainstreaming of Indian Microfinance". *Economic and Political Weekly*. XL.17(2005):1731-1739. Print.

Sen. Amartya. *Poverty and Famines: an Essay on Entitlement and Deprivation*.Oxford:OUP, 1981. Print.

—. *Commodities and Capabilities.* New Delhi: Oxford University Press 1999. Print.

—. "Gender and Cooperative Conflict". *Persistent Inequalities.* Ed. I. Tinker. New Delhi: OUP, 1990. Print.

—. *Capability and Well-Being, in the Quality of Life*. Ed. Martha. C. Nussbaum and A K. Sen. New York: Oxford University Press, 1993. Print.

—. *Development as Freedom*. Oxford: OUP, 1999. Print.

Smith, H.L. "Integrating Theory and research on the Institutional Determinants of fertility". *Demography.* 26.2 (1989): 171-184. Ed. Deepa Narayan *Measuring Empowerment: Cross Disciplinary Perspective*: 90, World Bank.

Tagore, Rabindranath. *Rabindra Rachanabali* (Revised edition), vol. 14. Kolkata: Visva Bharati, 1410.

Todaro, M. P. *Economic Development* (5th ed.). NewYork, London: Longman, 1994. Print.

World Bank. *Measuring Empowerment: Cross Disciplinary Perspective.* Ed. Deepa Narayan. Washington DC: World Bank, 1987. Print.

Yunus, M. *Banker to the Poor*. The U P Ltd. Dhaka, 1998. Print.

LIST OF CONTRIBUTORS

Anirban Bhattacharjee is a Junior Research Fellow in Cultural Studies at the Centre for Studies in Social Sciences, Calcutta & Jadavpur University, India. He has published widely in the fields of postcolonial studies, cultural studies, critical theory, Indian philosophy, and Nineteenth Century Bengal. He has delivered public talks at Jadavpur University, Hyderabad University, Delhi University, Jawaharlal Nehru University and the Indian Institute of Advanced Study, India and at 13th South Asian Literary Association (SALA) conference in Boston, USA. The proposed title of his doctoral thesis is "Philosophical Discourses of Housework in Bengal" and he is working under the supervision of Dr. Sibaji Bandyopadhyay, Professor of Cultural Studies at CSSS, Kolkata.

Bikas Chakraborti took his post-graduate degree from the University of Calcutta in 1970. He was awarded the Ph.D.degree on The Marxism of Che Guevara: An Analysis in 1992 by the same university. He authored books like Che Guevara: The Iconoclast (2000), Baul Jeevaner Samajttva (2003), Bharater Swadhinata Sangram (2007). He edited Folk, Folklore and Folkloristics: Interpenetration of Society and Folklore (2010).Also he contributed articles on folklore and sociology of music to many journals. Author was an Associate Professor of Political Science, Vidyasagar Evening College, Kolkata 700006.

Debalina Banerjee, is Assistant Professor of English, Vidyasagar Evening College. She is currently pursuing her Ph.D. on "'Mother' and 'Wife': Interrogating stereotypes in Five Indian English Women Novelists" from Visva-Bharati University. She has paid academic visits to Universities in the Netherlands, Spain, Germany, Austria and Czech Republic. She has co-edited a book with Kaustav Bakshi on *Indian Poetry in English* (2012) and another on *Nobel Laureates as Playwrights* (2013) with Dr. Benoy Banerjee. She is on the editorial board of the Journal *Annals of the University of Oradea* published by the faculty of Social-Humanistic Sciences, University of Oradea, Romania. Her areas of interest include women studies and culture studies.

Debasish Joddar has been teaching economics since 1999. At present, he is Assistant Professor, Department of Economics, Vidyasagar Evening

College under the University of Calcutta. He has also been pursuing research work as UGC teacher fellow on *Sustainability of Microenterprise and its Impact* under the Economics department of University of Kalyani. His areas of interest are rural economy, microfinance and gender related issues.

Hardik Brata Biswas received his initial training in Comparative Literature and switched to Women's Studies at Jadvapur University to do an M Phil on relations between visuality and gender in Bengal. Apart from being a PhD scholar at the same school working towards his dissertation on archives, everyday, image and the middle class women in Bengal, 1880s-1970s, he received a fellowship from SEPHIS under its History of Modernities and Sexualities in the Global South programme. At the School of Women's studies he collaborated with IFA to build a digital archive of Bengali women's images. He has written on popular photographic cultures (2011), transgression pornography and the middle-class Bengali family (2011) and edited a collection of 19th and early 20th century popular farces in Bengali: *Prohoshoney Kalikaler Bangamahila, 1860-1909* [Kaliyuga's Women in Farces, 1860-1909].

Margarete Rubik is Professor of English Literature at the University of Vienna, Austria, where she was Head of the Department several times. Her research interests are very wide, ranging from Restoration and eighteenth century literature to contemporary drama, and she has published widely in a variety of fields. She has written monographs on *Early Women Dramatists* and *The Novels of Margaret Oliphant*, and has edited the works of Delarivier Manley and Eliza Haywood, as well as essay collections, for instance, about *Aphra Behn and her Female Successors*, Jane Eyre adaptations (*A Breath of Fresh Eyre*), colonial fiction (*Stories of Empire*) and contemporary drama (*Staging Interculturality*).

Maysaa Jaber recently completed a PhD in English and American Studies from the University of Manchester, UK. Her doctoral work examined literary representations of criminal femme fatales in American hardboiled crime fiction in relation to medico-legal work on criminality, gender, and sexuality from the 1920s until the end of World War II. Now she is a lecturer at the University of Baghdad where she teaches different modules on literature to undergraduate students. She was a fellow in The University of Massachusetts Boston from September to November 2013. Her first manuscript *The Criminal Femme Fatales in American Hardboiled Crime Fiction* is forthcoming with Palgrave Macmillan.

M. Mohibul Haque is Assistant Professor in the Department of Political Science of Aligarh Muslim University, India. His areas of interest are: International Politics, Indian Polity, Minority Rights, Human Rights, Terrorism, Globalization, and Political Islam. He is a prolific writer and has contributed several articles, research papers, chapters in the edited books on socio-political issues of national and international importance. He has also authored a book (International Terrorism and Violence: A Human Rights Perspective). He has presented papers in national and international seminars/conferences in India and abroad. He has been participating in discussions on television channels and other public forums as panelist and discussant. Dr. M. Mohibul Haque is also the recipient of President of India Dr. Shankar Dayal Sharma Gold Medal for Academic Excellence.

Somdatta Mandal is Professor of English and current Chairperson at the Department of English and Other Modern European Languages, Visva-Bharati, Santiniketan. Her areas of interest are contemporary fiction, film and culture studies, Diaspora studies and translation. A recipient of several prestigious awards and fellowships like the Fulbright- Pre doctoral Fellowship, Fulbright Visiting Teaching Fellowship, Charles Wallace, Rockefeller Residency at Bellagio, Salzburg Seminar, Shastri Indo-Canadian Faculty Enrichment Fellowship, she has been published widely both nationally and internationally. She has written three academic books and edited and co-edited fifteen books and journals. She has also received a Sahitya Akademi award for translating short fiction.

Swati Ganguly is Associate Professor of English in the Department of English and Other Modern European Languages. She received the Charles Wallace Fellowship for Translation Studies in 1996 and has translated and co-edited (with Sarmistha Duttagupta) *The Stream Within: Short Stories by Contemporary Bengali Women* (1999). Swati is also the co-editor of the volume *Rabindranath Tagore and the Nation*. (2011) She has recently received The New India Foundation Fellowship to write a book on the history of Visva-Bharati (1951-1961). She writes fiction in Bengali.

Saoli Mitra is an eminent theatre personality of Bengal. She is at once a veteran actor, director and playwright. She received the Ibsen Centennial Award in January 2006. Her 'plays are one-woman performances, which look at the epic through a gender lens and throw a challenge to the audience.' Both her plays are based on the *Mahabharata*. The story of Draupadi is narrated in *Five Lords, Yet None a Protector (Nathabati Anathabat)*. The second, *Timeless Tales (Katha Amritasaman)*, is again set within the *Mahabharata*, looking at the 'destruction of an age and of a civilization'.

Samrat Sengupta is an Assistant Professor in the Dept. of English at Kharagpur College and also a Doctoral Scholar in Centre for Studies in Social Sciences, Calcutta. His area of research is Naxalbari Revolution and ethico-politics of postcolonial resistances in Bengal. He has co-edited an anthology of critical essays on postcolonialism and neocolonialism. He has written and published on Popular Culture, Film and Media, Deconstructive philosophy and Postcolonial theory. His latest work includes a special supplement on Violence and Terror in a Bengali Journal on Critical Theory.

Sreenanti Banerjee is a Junior Research Scholar and presently pursuing her MPhil in Social Sciences from Centre for Studies in Social Sciences, Calcutta (CSSSC). She had completed her Bachelor's degree in Sociology from St. Xavier's College (Autonomous), Calcutta and thereafter went to Tata Institute of Social Sciences, Mumbai (TISS) to pursue her Masters in Women's Studies. A Gold Medalist from TISS, Sreenanti's research interests are Feminist Theory, Continental Philosophy (especially Poststructuralist Theory), Cultural Studies, Postcolonialism, South Asian History and Sociology. Her present research is on Globalization, Feminist Politics and the Postcolonial Subject.

Sarmistha De has completed her PhD from Jadavpur University. Since 1994 she is an archivist in the State Archives of West Bengal. She is the author of *Marginal Europeans of Colonial India, 1860-1920,* (Thema, 2008) Her other major publications are: "The 'disposable' brethren: European marginals in eastern India during the Great Rebellion" in Biswamoy Pati Ed. *The Great Rebellion of 1857 in India: Exploring transgressions, contests, and diversities*, (London & New York: Routeledge, 2010), "Maidan: The Open Space in History", Sarmistha De and Bidisha Chakraborty, *Social Scientist*, Vol. 38, No 1-2, 2010, pp 3-22, "Marginal Whites and the Great Uprising of 1857: A Case Study of the Bengal Presidency" in Crispin Bates ed. *Mutiny at the Margins*, *Vol. II*, (Delhi Sage, 2013, Forth Coming), Sarmistha De & Bidisha Chakraborty, *19th century Calcutta: An Archival Exploration* (Delhi, Niyogi Books, 2013, forthcoming)

Sumana Gupta is Assistant Professor of English at Banipur Mahila Mahavidyalaya. Her areas of research and interest are Diaspora studies and the fiction of Bharati Mukherjee, Chitra Banerjee Divakaruni and

Jhumpa Lahiri. She has also published on Bharati Mukherjee's fiction in *Bharati Mukherjee: Critical Perspective*, edited by Dr. Somdatta Mandal.

Usha Bande, critic and academic, was till recently Fellow, Indian Institute of Advanced Study, Shimla and also Visiting Fellow at Visva Bharati University, Shanti Niketan (WB). Earlier she was on the faculty of English Literature in Govt. College for Women, Shimla. She retired as Principal Govt. College, Arki (Himachal Pradesh). Dr. Bande writes in English and Hindi and has translated Marathi stories into Hindi and English. Recipient of UGC Major Research Grant, she studied Hindi, Marathi, Punjabi and English short stories by women writers and brought out a critical book on the theme. She has numerous research papers and more than a dozen books to her credit which include: *Writing Resistance: A Comparative Study of Women Novelists; Violence in Literature, Media and TV; Ecology and Folk Traditions in Himachal Pradesh* and a Coffee Table book entitled *Forts and Palaces of Himachal Pradesh*. She is also associated with the Institute for Integrated Himalayan Studies, HP University, Shimla. Besides academic writing, Dr. Bande is fond of journalistic and creative writing and is a regular contributor to various magazines and news papers.

INDEX